# Management of Construction

Unlike the majority of construction project management textbooks out there, *Management of Construction Projects* takes a distinctive approach by setting itself in the context of a single and real-world construction project throughout and also by looking at construction project management from the constructor's perspective. This project-based learning approach emphasizes the skills, knowledge, and techniques students require to become successful project managers.

This second edition uses a brand new, larger, and more challenging case study to take students through key stages of the process, including:

- contracts and subcontracting;
- estimating, scheduling, and planning;
- supply chain and materials management;
- cost control, quality, and safety;
- project leadership and ethics; and
- claims, disputes, and project close-outs.

Also new to this edition is coverage of emergent industry trends such as LEAN, LEED, and BIM. The book contains essential features such as review questions, exercises, and chapter summaries, while example plans, schedules, contracts, and other documents are stored on a companion website (www.routledge.com/cw/schaufelbergerholm). Written in straightforward language from a constructor's perspective, this textbook gives a realistic overview and review of the roles of project managers and everything they need to know in order to see a successful project through from start to finish.

**John E. Schaufelberger** is the dean of the College of Built Environments at the University of Washington, USA.

**Len Holm** is a senior lecturer in construction management at the University of Washington, USA, and a construction management professional.

# Management of Construction Projects

## A Constructor's Perspective

Second edition

**John E. Schaufelberger and Len Holm**

Routledge
Taylor & Francis Group

LONDON AND NEW YORK

Second edition published 2017
by Routledge
2 Park Square, Milton Park, Abingdon, Oxon, OX14 4RN

and by Routledge
711 Third Avenue, New York, NY 10017

*Routledge is an imprint of the Taylor & Francis Group, an informa business*

© 2017 John E. Schaufelberger and Len Holm

First edition published by Prentice-Hall, 2002

*British Library Cataloguing-in-Publication Data*
A catalogue record for this book is available from the British Library

*Library of Congress Cataloging in Publication Data*
A catalog record for this data has been requested

ISBN: 978-1-138-69389-0 (hbk)
ISBN: 978-1-138-69391-3 (pbk)
ISBN: 978-1-315-52909-7 (ebk)

Typeset in Sabon
by HWA Text and Data Management, London
Printed and bound by CPI Group (UK) Ltd, Croydon, CR0 4YY

Visit the companion website: www.routledge.com/cw/schaufelbergerholm

# Contents

# Figures

# Tables

# Boxes

# Worksheets

# Preface

Successful construction projects are delivered by skilled project managers and superintendents. This book examines the skills, knowledge, tools, and techniques needed to be a successful project manager or superintendent from the perspective of the construction contractor's project team. The construction industry has become increasingly competitive, placing greater emphasis on effective construction project management, and many books have been written from the perspective of the owner's project manager. Few, however, have approached the subject from the contractor's perspective.

This book was developed for use as a text for undergraduate courses in construction project management and as a reference for construction contractors. It assumes that readers have a basic understanding of the construction process, the construction methods used in the industry, cost estimating, and project planning and scheduling. Topics are addressed just as a project manager and superintendent would in managing a construction project. The focus is on the individual management processes and techniques needed to manage a project, and tools are provided to assist in the performance of these processes. While the context for the discussion is management of commercial projects, the principles and techniques presented also are applicable to residential, industrial, and heavy construction projects.

Each chapter has a similar organization. Topics are first discussed in general terms, then individual issues are discussed in detail and illustrated. A single construction project is used throughout the book, providing a context for concept illustration and student exercises. While the construction company used in this text is fictitious, the project was actually constructed in Seattle, Washington. Construction progress photographs and contract documents are shown on the companion website. Forms illustrated are shown with project-specific information entered in italics. The chapters conclude with a set of review questions that emphasize the major points covered in the chapter, and an instructor's manual containing answers to the review questions is available on the companion website. Exercises are provided to allow students to apply the principles learned. A listing of all abbreviations used in the text and a glossary of construction terms are included.

This second edition features updated chapters, a new, more complex example project, a new chapter on preconstruction planning, a new chapter on business aspects of project management, and a new chapter on construction project leadership. Included are coverage of contemporary project management topics such as use of building information models, sustainable construction, environmental compliance, lean construction, and off-site construction.

This book could not have been written without the help of many people. We acknowledge the following: the University of Washington for allowing us to use their building and photographs of what turned out to be an excellent example of a team-built project; Bob Vincent, project manager, and Jesse Hallowell, project superintendent, for providing access to project management documents; ConsensusDocs for granting us permission to reproduce their contract forms; and two former students, Molly Roe and Avra Platis, for their assistance in obtaining project documentation.

*John Schaufelberger,*
*Len Holm*

# Companion website

The following documents are available on the companion website, www.routledge.com/cw/schaufelbergerholm

The following are available to students and instructors:

- ConsensusDocs® contracts referred to within the text:
  - Form 500:     Owner-CM-At-Risk
  - Form 500.1:    Guaranteed Maximum Price Amendment
  - Form 751:     Subcontract Agreement (Chapter 9)
  - Form 814:     Certificate of Substantial Completion (Chapter 11)
- Expanded versions of some figures utilized within the text:
  - Figure 3.5L:    Project Item List
  - Figure 3.14L:   General Conditions Estimate
  - Figure 3.15L:   Summary Estimate
  - Figure 4.3L:    Cost Loaded Schedule
  - Figure 7.2L:    Long Form Purchase Order
  - Figure 9.2L:    Documents Exhibit
- Other documents referred to within the text:
  - Detailed Construction Schedule (Chapter 4)
  - Jobsite Layout Plan (Chapter 8)
- Select drawings from our case study project: 3 files, 52 sheets
- Select specification sections from our case study project: 2 files, 26 spec sections
- 50 photographs of our case study project while under construction
- Live Excel estimating forms utilized in Chapter 3 of this text and in *Construction Cost Estimating, Process and Practices,* Pearson, 2005

The following are also available to instructors:

- Instructor's Manual, complete with answers to all of the review questions and some of the exercises. A select group of case studies from *Who Done It? 101 Case Studies in Construction Management* (Amazon/Create Space, 2015) are also included.
- PowerPoint lecture slides for all 19 chapters, 19 separate files, over 560 slides in total.

# Abbreviations

| | |
|---|---|
| AAA | American Arbitration Association |
| ACT | acoustical ceiling tile |
| ADR | alternative dispute resolution |
| AGC | Associated General Contractors of America |
| AHJ | authority having jurisdiction, often city building department |
| AIA | American Institute of Architects |
| ASI | architect's supplemental instructions |
| ASTM | American Society for Testing and Materials |
| ATB | asphalt treated base |
| BIM | building information models, or modeling |
| BOT | built-operate-transfer |
| CAD | computer-aided design |
| CCA | construction change authorization |
| CCD | construction change directive |
| CCE | *Construction Cost Estimating, Process and Practices* (text) |
| CD | construction documents, also change directive |
| CEO | chief executive officer |
| CIP | cast in place (concrete) |
| CM | construction manager or management |
| CM/GC | Construction Manager/General Contractor, also GC/CM |

| CMU | concrete masonry unit |
|---|---|
| CO | change order |
| C of O | certificate of occupancy |
| COP | change order proposal |
| CPFF | cost-plus-fixed-fee |
| CPPF | cost plus percentage fee |
| CSF | 100 square feet, also SQ or square |
| CSI | Construction Specifications Institute |
| CY | cubic yard |
| DBIA | Design Build Institute of America |
| DBOM | design-build-operate-maintain |
| DD | design development |
| DRB | dispute resolution board |
| EA | each |
| EJCDC | Engineers Joint Council Document Committee |
| EMR | experience modification rating |
| EMT | electrical metallic conduit, thin wall |
| E & O | errors and omissions (insurance) |
| FE | field engineer |
| FICA | Federal Insurance Contributions Act (Social Security) |
| FOB | free on board, also freight on board |
| GBI | Green Building Institute |
| GC | general conditions, also general contractor |
| GMP | guaranteed maximum price |
| GWB | gypsum wall board, also sheetrock or drywall |
| HVAC | heating, ventilating, and air conditioning |
| IDC | interim directed change |
| IPD | integrated project delivery |
| ITB | instructions to bidders, also invitation to bid |
| LDs | liquidated damages |
| LEED | Leadership in Energy and Environmental Design |
| LF | linear feet |
| LOI | letter of intent |
| LS | lump sum |
| MEP | mechanical, electrical, and plumbing (systems and contractors) |

| MH | man-hour |
| M & M | means and methods |
| MOCP | *Management of Construction Projects, a Constructor's Perspective* (text) |
| NA | not applicable |
| NAHB | National Association of Home Builders |
| NIC | not-in-contract |
| NTP | notice to proceed |
| NWCC | Northwest Construction Company |
| OAC | owner architect contractor (meeting) |
| OH & P | overhead and profit, also known as fee |
| OIC | officer-in-charge |
| O & M | operation and maintenance (manual) |
| OM | order-of-magnitude (estimate) |
| OSHA | Occupational Safety and Health Administration |
| PE | project engineer, also pay estimate |
| PM | project manager or management |
| PO | purchase order |
| PPE | personal protection equipment |
| Precon | preconstruction (phase or agreement) |
| PSI | pounds per square inch |
| PVC | polyvinyl chloride |
| PPP | public–private partnerships |
| QC | quality control |
| QTO | quantity take-off |
| QTY | quantity |
| Rebar | concrete reinforcement steel |
| Recap | cost recapitulation sheet (estimating) |
| RFI | request for information, or interpretation |
| RFP | request for proposal |
| RFQ | request for qualifications, also request for quotation |
| ROM | rough-order-of-magnitude |
| ROT | rule-of-thumb |
| SDS | safety data sheets |
| SF | square feet |
| SFCA | square feet of contact area |

| | |
|---|---|
| SK | sketch |
| SOG | slab on grade |
| SOV | schedule of values (pay estimate) |
| Spec(s) | specifications |
| SPM | senior project manager |
| STP | Supervisory Training Programs (AGC) |
| Sub | subcontractor |
| Super or Supt | superintendent |
| SWPPP | storm water pollution prevention plan |
| TCO | temporary certificate of occupancy |
| TCY | truck or loose cubic yards |
| T & M | time and materials (contract or billing) |
| UMH | unit man-hours |
| UP | unit price |
| USGBC | United States Green Building Council |
| UW | University of Washington |
| VE | value engineering |
| WBS | work breakdown structure |
| ZGF | Zimmer Gunsul Frasca (Architects) |

# 1 | Introduction

## Project management concepts

Project management is the application of knowledge, skills, tools, and techniques to the many activities required to complete a project successfully. In construction, project success generally is defined in terms of document control, safety, quality, cost, and schedule. These project attributes can be visualized as depicted in Figure 1.1. The project manager's challenge is to balance quality, cost, and schedule within the context of a safe project environment while maintaining control of the many construction documents. While cost and schedule may be compromised to produce a quality project, there can be no compromising regarding safety, and proper documentation is required to ensure compliance with contract requirements.

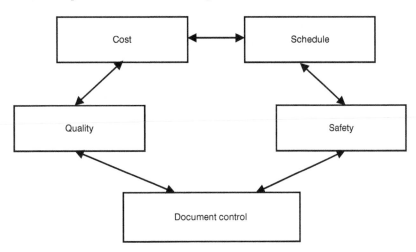

**Figure 1.1** Critical project attributes

In this book, we will examine project management from the perspective of the construction contractor. Other project managers typically are involved in a project representing the owner and the designer, but our focus is on the knowledge, skills, tools, and techniques needed to be successful as a project manager for a construction contractor. Our context will be that of a project manager for a commercial general contractor. The principles and techniques discussed, however, are equally applicable on residential, specialty, industrial, and infrastructure or heavy construction projects.

The project manager is the leader of the contractor's project team and is responsible for identifying project requirements and leading the team in ensuring that all are accomplished safely and within the desired budget and time frame. To accomplish this challenging task, the project manager must organize his or her project team, establish a project management system that monitors project execution, and resolve issues that arise during project execution. In successive chapters, we will discuss the many tools that a project manager and superintendent should use in managing a project. Not all may apply to every project, but the project manager and superintendent must select those that are applicable for each project.

## The major phases of a construction project

Project planning: involves evaluation of the risks that are associated with the project, particularly those related to safety, cost, quality, or schedule. Risk analysis and risk management are critical skills essential to successful project management. The project manager develops the organizational structure needed to manage the project and the communications strategy to be used within the project management organization and with other project stakeholders. Material procurement and subcontracting strategies also are developed during the planning phase. Topics relating to project planning are discussed in Chapters 3 through 7.

Project start-up: involves mobilizing the project management team, educating them regarding the project and associated risks, and conducting team-building activities. The project management office is established, and project documentation management systems are created. Initial project submittals are provided to the owner or the owner's representative. Vender accounts are established, and materials and subcontract procurement are initiated. Project cost, schedule, and quality control systems are established to manage project execution. Topics relating to project start-up are addressed in Chapters 8 and 9.

Project control: involves controlling the project, interfacing with external members of the project team, anticipating risk by taking measures to mitigate its potential impacts, and adjusting the project schedule to accommodate changing conditions. The project manager monitors the document management system, the quality management system, cost control system, and schedule control system, making adjustments where appropriate. He or she reviews performance reports to look for variances from expected performance and takes action to minimize their impacts. Topics relating to project control are addressed in Chapters 10 through 16.

Project close-out: involves completing the physical construction of the project, submitting all required documentation to the owner, and financially closing out the project. The project manager must pay close attention to detail and motivate the project team to close out the project expeditiously to minimize overhead costs. Project close-out is discussed in Chapter 17.

Post-project analysis: involves reviewing all aspects of the project to determine lessons that can be applied to future projects. Such issues as anticipated cost versus actual cost, anticipated schedule versus actual schedule, quality control, subcontractor performance, material supplier and construction equipment performance, effectiveness of communications systems, and work force productivity should be analyzed. Many contractors skip this phase, and simply go to the next project. Those who conduct post-project analyses learn from their experiences and continually improve their procedures and techniques. Post-project analysis is discussed in Chapter 17.

## Project delivery methods

The principal participants in any construction project are the owner or client, the designer (architect or engineer), and the general contractor. The relationships among these participants are defined by the project delivery method used for the project. The choice of delivery method is the owner's, but it has an impact on the scope of responsibility of the contractor's project manager. Owners typically select project delivery methods based on the amount of risk that they are willing to assume and the size and experience of their own contract management staffs. In this section, we will examine the seven most common project delivery methods that are used in the United States.

## Traditional project delivery method

The traditional project delivery method is illustrated in Figure 1.2. The owner has separate contracts with both the designer and the general contractor. There is no contractual relationship between the designer and the general contractor, but they must communicate with each other throughout the construction of the project. Typically, the design is completed before the contractor is hired in this delivery method. The contractor's project manager is responsible for obtaining the project plans and specifications, developing a cost estimate and project schedule for construction, establishing a project management system to manage the construction activities, and managing the construction.

Sometimes the project owner has sufficient construction management staff to manage the construction without benefit of a general contractor and chooses to contract directly with all of the specialty contractors. This is known as the multiple prime project delivery method and is a variation on the traditional project delivery method. This approach is illustrated in Figure 1.3. In this delivery method, the owner's project manager creates the project schedule and manages the work of the specialty contractors.

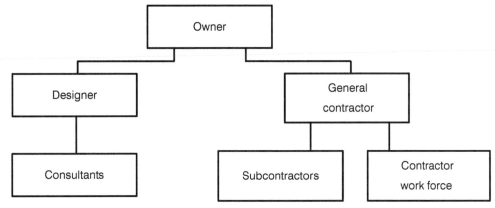

**Figure 1.2**   Traditional project delivery method

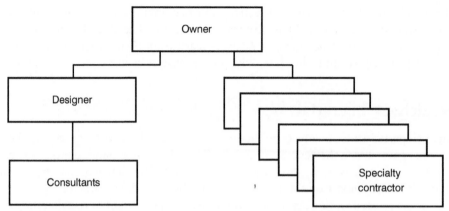

**Figure 1.3**   Traditional project delivery method with multiple prime contractors

## Agency construction management project delivery method

In the agency construction management project delivery method, the owner has three separate contracts (one with the designer, one with the general contractor, and one with the construction manager) as illustrated in Figure 1.4. The construction manager acts as the owner's agent (also known as the owner's representative) and coordinates design and construction issues with the designer and the general contractor. The construction manager usually is the first contract awarded, and he or she is involved in hiring both the designer and the general contractor. In this delivery method, the general contractor usually is not hired until the design is completed. The general contractor's project manager has responsibilities in this delivery method similar to those listed for the traditional project delivery method. The primary difference is that the general contractor's project manager interfaces with the construction manager in this delivery method instead of with the owner, as is the case in the traditional method.

Another form of the agency construction manager project delivery method is shown in Figure 1.5. In this project delivery method, there is no single general contractor. Instead, the project owner awards separate contracts to multiple specialty contractors and then tasks the construction

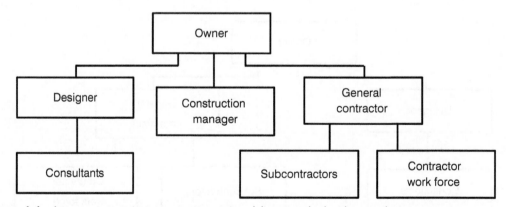

**Figure 1.4**   Agency construction management project delivery method with general contractor

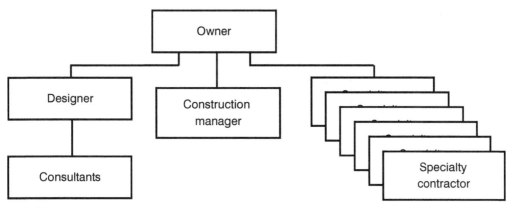

**Figure 1.5** Agency construction manager project delivery method with multiple prime contractors

manager to coordinate their activities. The construction manager has no contractual relationship with any of the specialty contractors and performs none of the construction tasks nor procures any materials for the project.

## Construction manager-at-risk project delivery method

In the construction manager-at-risk project delivery method, the owner has two contracts (one with the designer and one with the construction manager) as illustrated in Figure 1.6. In this delivery method, the designer usually is hired first. The construction manager typically is hired early in the design development to perform a variety of preconstruction services, such as cost estimating, constructability analysis, and value engineering studies. Once the design is completed, the construction manager becomes the general contractor and constructs the project. Some project owners require the construction manager to subcontract all of the work to specialty contractors while other owners allow the construction manager to self-perform selected scopes of work. In some cases, project construction may be initiated before the entire design is completed. This is known as fast-track or phased construction. The contractor's project manager interfaces with

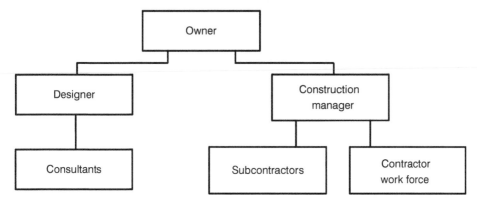

**Figure 1.6** Construction manager-at-risk project delivery method

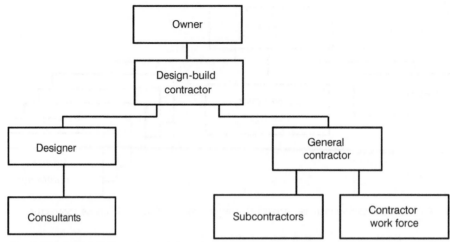

**Figure 1.7**    Design-build project delivery method

the designer and manages the execution of preconstruction tasks. Once construction starts, the project manager's responsibilities are similar to those in the traditional project delivery method. This is the project delivery method that was used in the project that serves as the case study used throughout this book.

## Design-build project delivery method

In the design-build project delivery method, the owner has a single contract with the design-build contractor for both the design and construction of the project as illustrated in Figure 1.7. The design-build contractor may have a design capability within its own organization, may chose to enter into a joint venture with a design firm, or may contract with a design firm to develop the design. On some projects, a design firm may sign the contract and hire the construction firm. Construction may be initiated early in the design process using fast-track procedures or may wait until the design is completed. In this delivery method, the contractor's project manager is responsible for interfacing with the owner and managing both the design and the construction of the project. A variation to this method is the design-build-operate-maintain (DBOM) project delivery method in which the contractor operates the facility after construction for a specified period for an annual fee. This delivery method often is used for utilities projects such as water treatment or sewage treatment plants.

## Bridging project delivery method

The bridging project delivery method is a hybrid of the traditional and the design-build delivery methods. The owner contracts with a design firm for the preparation of partial design documents. These documents typically define functional layout and appearance requirements. A design-build contractor is then selected by the owner to complete the design and construct the project.

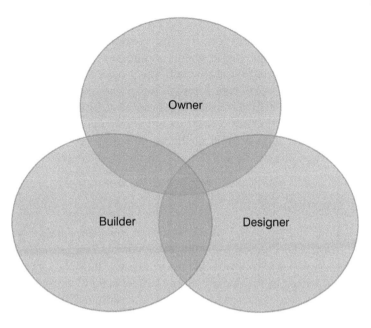

Collaborative Delivery

**Figure 1.8**   Integrated project delivery method

# Integrated project delivery method

The design and construction contracts awarded in the other project delivery methods are transactional contracts, written for the purpose of acquiring design and construction services. The contract awarded in the integrated project delivery (IPD) method is a relational contract that creates a collaborative relationship among the owner, designer, and builder as illustrated in Figure 1.8. An example of this type of contract is ConsensusDocs 300, Standard Multi-Party Integrated Project Delivery Agreement.

Under the contract provisions, the three parties commit to collaborating in the design, construction, and commissioning of the project and agree to share in the risks and rewards associated with the project. Typically, there is a target project cost and a risk pool, similar to a contingency, established in the contract to address project risks and cover unanticipated expenses. When the project is completed, any funds remaining in the risk pool are equally distributed among the three parties. In this project delivery method, the project is managed by a management group composed of senior representatives of the owner, designer, and builder who are empowered to serve as the decision-making body for delivery of the project, and there is total fiscal transparency among the three representatives.

# Public–private partnership project delivery method

Contracts awarded in the public–private partnership (PPP) project delivery method are known as build-operate-transfer (BOT) contracts. Under these contracts, a private sponsor, usually a consortium or joint venture, executes a contract with a public agency to finance, design, construct, and operate a project for a specified period of time. At the end of the operating or concession period, ownership of the project is transferred from the private sponsor to the government agency. In this delivery method, a government agency identifies project requirements, establishes the concession period for project operation, solicits proposals, and awards the contract. The project designer and constructor may be joint venture participants or may be subcontractors to the private sponsor. Funding for the project may come from equity participation in the sponsoring consortium, loans, or the sale of bonds. During the concession period, the project sponsor collects revenues from operation of the project to recover its investment and earn a profit. This delivery method is often used for large infrastructure projects such as highways that charge tolls for use and power plants that sell generated electricity.

## Sustainable construction

The design, construction, and operation of projects have environmental, economic, and social impacts. Sustainable construction means evaluating these potential impacts and selecting strategies that emphasize the positive impacts while minimizing the negative impacts. The overall sustainability of a project is greatly influenced by the design, but the builder can influence the environmental and social impact of the construction operations. Some contractor sustainable actions include:

- reuse of materials and minimization of construction waste;
- minimization of noise, light, and air pollution during construction;
- protection and restoration of the natural environment;
- elimination of storm water runoff and soil erosion; and
- selection of construction materials with high recycle content.

Buildings that are designed and constructed to achieve sustainability goals are often referred to as "green" buildings. There are several systems that have been developed to provide a framework for assessing the achievement of sustainability goals.

The United States Green Building Council (USGBC) (www.usgbc.org) developed its Leadership in Energy and Environmental Design (LEED) green building rating system to provide a common standard for the measurement of what constitutes a high-performance green building. The LEED rating system has been revised several times, but it provides for four levels of certification and measures compliance with eight categories of standards. The four levels of certification are:

- certified,
- silver,
- gold, and
- platinum.

The eight categories of standards are

- location and transportation,
- sustainable sites,
- water efficiency,
- energy and atmosphere,

- materials and resources,
- indoor environmental quality,
- innovation, and
- regional priority.

The level of certification achieved by a project depends on the number of credits awarded by a third-party certification process. This process is described in more detail in Chapter 5.

Another assessment process is known as Green Globes, which is distributed and run by the Green Building Institute (GBI) (www.thegbi.org). The Green Globes rating system includes a web-based self-assessment tool, a rating system for certification of a building, and a guide for enhancing the sustainability of a project. Its ratings are based on project management; site; energy; water; resources, building materials, and solid waste; emissions and effluents; and indoor environment. Based on verification by a third-party, certification can be granted as one to four globes, with four globes being the highest. Certification levels are based on the percentage of applicable points achieved by the project.

# Building information models

A building information model is an integrated database containing parametric information and design documents for the entire building or other project. The data are used to create three-dimensional projections of the building as illustrated in Figure 1.9 or may be used to depict the locations of underground utility lines or structures. Building information modeling (BIM) involves the process of designing a building collaboratively using a coherent system of computer models. The architectural, structural, mechanical, electrical, and plumbing models can be brought together to discover any clashes or elements of the separate models that occupy the same space. BIM can be used throughout the design and construction processes to illustrate the designers' intent, to simulate construction sequences, and to manage construction activities. It can be used to develop quantity take-offs of building components to support development of cost estimates, to develop site logistics plans, to create construction schedules, and to identify opportunities for offsite prefabrication of building components. The use of BIM for preconstruction planning will be discussed in more detail in Chapter 5.

# Lean construction

Lean construction is a continuous process for analyzing the delivery of construction projects to eliminate waste while meeting or exceeding project owner expectations. The application of lean principles results in better utilization of resources, especially labor and materials. While lean construction does not replace the project master schedule, it does utilize better short-interval planning and control that improve the timely completion of project tasks. The strategy for lean supply is to provide materials when needed to reduce variation, eliminate waste, improve workflow, and increase coordination among the trades. The project manager's responsibility is to create a realistic construction flow that shows the contractor's and the subcontractors' dependency on material suppliers. Material deliveries are then scheduled so that the materials arrive on site just as they are needed for installation on the project. Another aspect of lean construction is the expanded use of off-site construction or

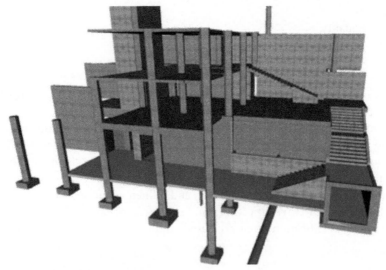

**Figure 1.9**    Building information model

prefabrication of building components. Reduced fabrication on the project site enables workers to concentrate on component installation rather than construction, saving time, reducing material waste, enhancing project safety, and enhancing quality. Examples of lean construction techniques are discussed in more detail in Chapter 5. Accomplishment of lean construction objectives requires a cultural change in the behavior of the project team including the subcontractors.

## Project management organizations

The size and structure of the project management organization depend on the size of the project, its complexity, and its location with respect to other projects or the contractor's home office. The cost of the project management organization is considered project overhead and must be kept economical to ensure the cost of the contractor's construction operation is competitive with that of other contractors. The goal in developing a project management organization is to create the minimum organization needed to manage the project effectively. If the project is unusually complex, it may require more technical people than would be required for a simpler project. If the project is located near other projects or the contractor's home office, technical personnel can be shared among projects, or backup support can be provided from the home office. If the project is located far from other contractor activities, it must be self-sufficient.

General contractors typically organize their project management teams in one of three organizational structures. In one type of structure, estimating and scheduling are performed in the contractor's home office as illustrated in Figure 1.10. In an alternative organizational structure, estimating and scheduling are the project manager's responsibilities, as illustrated in Figure 1.11. Both the project manager and the superintendent may report to the contractor's chief of construction operations as illustrated in Figure 1.11. In the third alternative organizational structure, the superintendent may report to the project manager, as illustrated in Figure 1.12. The choice of project management organizational structure depends on the contractor's approach to

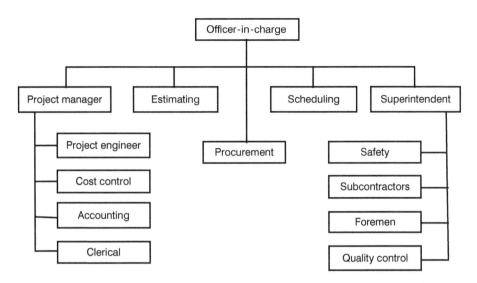

**Figure 1.10**   Contractor's project management organization with estimating and scheduling at home office

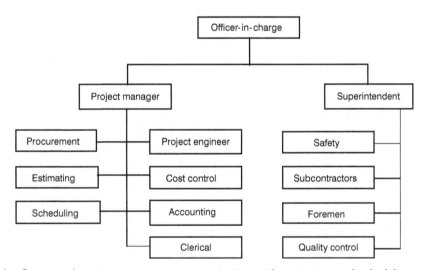

**Figure 1.11**   Contractor's project management organization with estimating and scheduling under project manager

managing projects. The officer-in-charge (OIC) is the project manager's supervisor. He or she may have various titles as described later in this chapter.

## Project team development

Once the project management organization is selected for a project, the project manager identifies the individuals to be assigned to each position. Most people will come from within the contractor's

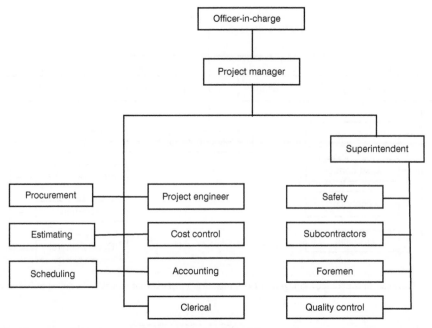

**Figure 1.12** Contractor's project management organization with superintendent reporting to project manager

organization, but some may be hired externally. Selection of project team members from inside the construction firm may be made by the project manager or may be made by senior company managers. If new people are hired, the project manager must prepare a job description for each position and a list of skills needed to perform the job, recruit and select new employees, and train them on the contractor's methods of doing business. These human resource management responsibilities are explained in more detail in Chapter 18. Once all of the team members have been selected, the project manager must forge them into a cohesive team. This requires team building, which may be a significant challenge for the project manager. Team building is discussed in Chapter 5.

## Project team member responsibilities

Individual team member responsibilities may vary from company to company and from project to project, but in general they are as described below. The superintendent is the only position that is specified in most construction contracts.

### Officer-in-charge

The OIC is the principal official within the construction company who is responsible for construction operations. He or she generally signs the construction contract and is the individual to whom the owner turns in the event of any problems with the project manager. This individual often carries the title of vice president for operations, operations manager, district manager, or may be the owner or chief executive officer (CEO).

## Senior project manager

In some construction companies, senior project managers are appointed to oversee multiple projects. When they are used, senior project managers are between the OIC and the project managers.

## Project manager

The project manager has overall responsibility for completing the project in conformance with all contract requirements within budget and on time. He or she organizes and manages the contractor's project team. Specific responsibilities include:

- coordinating and participating in the development of the project budget and schedule;
- developing a strategy for executing the project in terms of what work to subcontract;
- communicating frequently with the owner and the designer;
- soliciting, issuing contract packages, evaluating, and awarding subcontracts and material purchase orders;
- negotiating and finalizing contract change orders with the owner and subcontractors;
- implementing project cost and schedule control procedures;
- scheduling and managing project team meetings;
- supervising project office staff;
- submitting monthly progress payment requests to the owner; and
- managing project close-out activities.

## General superintendent

In some construction companies, a general superintendent may be appointed for a single large complex project, with area superintendents selected for major areas within the project. In other companies, the general superintendent is in charge of all superintendents employed by the company.

## Superintendent

The superintendent is responsible for the direct daily supervision of construction activities on the project, whether the work is performed by the contractor's employees or those employed by subcontractors. Specific responsibilities include:

- planning, scheduling, and coordinating the daily activities of all craft labor working on the site;
- determining the construction building methods and work strategies for work performed by the contractor's own work force;
- preparing a 3-week look-ahead schedule and coordinating equipment requirements and material deliveries;
- preparing daily reports of project activities;
- submitting daily time cards for self-performed work;
- ensuring all work performed conforms to contract requirements; and
- ensuring all construction activities are conducted safely.

## Project engineer or field engineer

The project or field engineer is responsible for resolving any technical issues relating to completion of the project. On small projects, the project engineer's responsibilities may be performed by the project manager. Specific responsibilities include:

- maintaining submittal and request for information logs;
- reviewing submittals and transmitting them to the owner or designer;
- preparing and submitting requests for information;
- preparing contract documents and correspondence and maintaining the contract file; and
- reviewing subcontractor invoices and requests for payment.

In some companies, this position might be called *assistant project manager*.

## Quality control inspector

On some projects, the owner may require or the contractor may choose to assign a quality control inspector. This individual continually inspects the work under construction to ensure compliance with contract quality requirements.

## Safety inspector

On some projects, the contractor may choose to assign a full-time safety inspector to ensure that all required safety equipment is worn and all accident prevention measures are taken by all individuals working on site, whether they are employed by the contractor or anyone else.

## General foreman and assistant superintendent

On some projects, the contractor may choose to have general foremen who are responsible for all the work being performed in selected areas of the project. In a similar manner, some contractors may assign assistant superintendents to assist the superintendent in the performance of his or her responsibilities.

## Foremen

The foremen are responsible for the direct supervision of the craft labor on the project. The construction firm will assign foremen for work that is performed by the company's own construction crews. Foremen for all subcontracted work will be assigned by each subcontractor. Specific responsibilities include:

- coordinating the layout and execution of individual trade work on the project site;
- verifying that all required tools, equipment, and materials are available on site;
- ensuring all craft work conforms to contract requirements; and
- preparing daily time sheets for all supervised personnel.

## Project management ethics

A construction company is judged by the integrity that it demonstrates in conducting its business and in treating its employees. A key component of the company's reputation is the ethics of the company leaders and employees. Ethics are the moral standards used by people in making personal and business decisions. They involve determining what is right in a given situation and then having the courage to do what is right. Each decision has consequences, often to ourselves as well as to others. Generally, there are three primary ethical directives: loyalty, honestly, and responsibility. Loyalty means doing what is in the best interest of the construction company and the project owner. Honesty is more than truth telling; it involves the correct representation of ourselves, our actions, and our words. Responsibility involves anticipating the consequences of our actions and taking responsible measures to prevent harmful occurrences. While laws and regulations define courses of action with which we must legally comply, ethics are standards of conduct that help us make the right decisions.

A construction company's ability to acquire and maintain customers will be greatly influenced by the potential customer's perception of the ethics of the construction firm and its project managers. The project manager establishes the ethical culture for the entire project team and leads by example. Additional discussion on project leadership is contained in Chapter 19. Subcontractors, material suppliers, and project owners may choose not to do business with project managers who have reputations for unethical behavior. The following paragraphs address the most common ethical challenges to be faced in obtaining a construction contract and performing the work needed to complete the project.

## Bid shopping

Bid shopping occurs when general contractors disclose to prospective subcontractors the price quotation received from competing subcontractors. The intent is to encourage subcontractors to lower their quotations. Such practice is considered unethical. The subcontractors are being asked to submit their best prices for a specific scope of work, and they provide their quotations to the general contractor with the expectation that they will be kept confidential and not shared with competitors. Another form of bid shopping that is unethical occurs when a general contractor uses the quotation from one subcontractor in their bid but selects another subcontractor to perform the work.

## Timely payment of subcontractors and suppliers

Subcontracts and purchase orders (supply contracts) often contain a provision that the general contractor will make payment once they receive the monthly payment from the project owner. It is unethical for the general contractor to withhold payment to the subcontractors and suppliers for an extended period of time after receipt of the payment from the owner. While withholding the payments may assist the general contractor in managing its cash flow, it has a significant adverse effect on the financial condition of the subcontractors and suppliers. Construction companies with reputations for late payment of subcontractors and suppliers often have difficulty obtaining competitive quotations from subcontractors and suppliers.

## Withholding information

Situations may arise on a project that may impact subcontractors. This may include changes in the construction schedule or differing site conditions. The project manager needs to ensure that all parties potentially impacted by the information are given timely notice.

The best strategy with respect to ethical behavior on a project is to adhere to the so-called Golden Rule, which is to treat others as we would wish to be treated.

## Summary

The contractor's project manager is the leader of the contractor's project management team. He or she is responsible for managing all the activities required to complete the job on time, within budget, and in conformance with quality requirements specified in the contract. The major phases of a construction project are: project planning, project start-up, project control, project close-out, and post-project analysis.

There are seven major project delivery methods used in the United States. The primary differences among them are the relationships between the project participants. In the traditional delivery method, the owner has separate contracts with the designer and the general contractor. There is no contractual relationship between the designer and the general contractor. In the agency construction management delivery method, the owner has separate contracts with the construction manager, designer, and the general contractor. The construction manager acts as the owner's representative on the project but has no contractual relationship with either the designer or the general contractor. In the construction management-at-risk delivery method, the owner has two separate contracts, one with the designer and one with the general contractor who also acts as the construction manager. In the design-build delivery method, the owner has a single contract with the design-build contractor who is responsible for both designing and constructing the project. The bridging delivery method is a hybrid method in which the owner first contracts with a designer for a partial design and then contracts with a design-build firm to complete the design and construct the project. In the IPD method, the owner, designer, and builder form a collaborative relationship to execute the project. In the PPP project delivery method, a private sponsor finances, designs, constructs, and operates the project.

Sustainable construction involves the design and construction of projects that minimize adverse environmental and social impacts of construction. Building information models are integrated databases that can be used to illustrate the designer's intent, to simulate construction sequences, and to manage construction activities. Lean construction involves the selection of processes for delivering construction projects to eliminate waste.

Contractors establish project management organizations to manage construction activities. The project team typically consists of a project manager, superintendent, project engineer, foremen for self-performed work, and administrative support personnel depending upon project size and complexity. Project management ethics are moral standards used by project team members in making personal and business decisions. The project manager establishes the ethical culture for the project team and leads by example.

## Review questions

1. What are five critical project attributes that the project manager must integrate?
2. What are the major phases of a construction project, and what occurs during each phase?
3. What is the difference between the traditional and the construction management delivery methods?
4. What is the difference between the construction manager-at-risk and the design-build delivery methods?
5. What is the difference between the design-build delivery method and the IPD method?
6. How do the responsibilities of the project manager differ from those of the project superintendent?
7. What are the major duties of the project or field engineer?
8. What is bid shopping, and why is it considered to be unethical behavior?

## Exercises

1. Develop an organization chart for a project management organization to manage the construction of a $20 million office complex that is to be completed within 1 year.
2. Describe the advantages and disadvantages of the project management organization shown in Figure 1.12 as compared with the organization shown in Figure 1.11.
3. Redraw Figure 1.4 depicting the lines of communication among the project participants.

# 2

# Construction contracts

## Introduction

A contract is a legal document that describes the rights and responsibilities of the parties, for example, the owner and the general contractor. The terms and conditions of their relationship are defined solely within the contract documents. These documents should be read and completely understood by the contractor before deciding to pursue a project. They also are the basis for determining a project budget and schedule. To manage a project successfully, the project manager must understand the organization of the contract documents and understand the contractual requirements for his or her project. This knowledge is essential if the project manager has any expectation of satisfying all contractual requirements.

The contract documents describe the completed project and the terms and conditions of the contractual relationship between the owner and the contractor. Usually there is no description of the sequence of work or the means and methods to be used by the contractor in completing the project. The contractor is expected to have the professional expertise required to understand the contract documents and select appropriate subcontractors or qualified tradespeople, materials, and equipment to complete the project safely and achieve the quality requirements specified. For example, the contract documents will specify the dimensions and workmanship requirements for cast-in-place concrete structural elements but will not provide the design for required formwork.

## Organization of contract documents

A typical construction contract consists of the five documents shown in Figure 2.1. A special conditions section is not always used, unless there are some specific requirements or restrictions placed on the contractor for the project. In some cases, such as the ConsensusDocs example used in this book, the agreement and general conditions are combined into a single document. Some project owners use a project manual that contains the general conditions, special conditions, soils report, erosion control requirements, prevailing wage rates, and other project related documents.

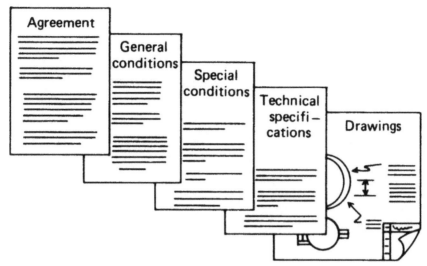

**Figure 2.1**   Contract documents

The invitation to bid and request for proposals generally are not considered contract documents. Some public owners issue a public notice to bidders providing a brief description of the project, the method for obtaining the bid package, as well as the date, time, and location for submission of bids. Other owners call their solicitation documents instructions to bidders or instructions for bidders.

Any of the contract documents can be modified by addenda or by change orders. Addenda are issued by the owner prior to award of the contract, and change orders are executed between the owner and the contractor after contract award. Any addenda or change orders used also are considered to be contract documents.

The agreement describes the project to be constructed; the pricing method to be used and cost, if lump sum, or the fee structure, if cost-plus; the time allowed for construction; any liquidated damages; and the name of the owner and the contractor. Supporting documents may be incorporated into the agreement by reference or as exhibits. Examples might be a geotechnical evaluation of the project site, erosion control plan requirements, traffic control plan requirements, and hazardous materials management plan requirements. These documents then become part of the contract documents.

The general conditions provide a set of operating procedures that the owner typically uses on all projects. They describe the relationship between the owner and the contractor, the authorities of the owner's representatives or agents, and the terms of the contract. Some owners use standard general conditions published by professional organizations such as those discussed later in this chapter. Topics typically addressed in the general conditions are:

- rights and responsibilities of the owner;
- authorities of the designer or construction manager;
- rights and responsibilities of the contractor;
- bonds and insurance;
- changes in contract scope of work;
- changes in contract price or fee;

- changes in contract time;
- inspections;
- uncovering defects and correction of work;
- protection of personnel and property;
- progress payments;
- subcontracting;
- contract close-out procedures;
- warranty of work;
- resolution of disputes; and
- early termination of the contract.

The special or supplementary conditions contain any unique requirements for the specific project. Examples of special conditions requirements are:

- site access restrictions;
- site security requirements;
- permission requirements for night or weekend work;
- location of parking for contractor's employees;
- listing of equipment furnished by owner;
- required insurance coverage limits; and
- mandatory (prevailing) wage rates (generally only on public projects).

Owners who use standard general conditions also may use the special conditions to modify selected general conditions requirements for a specific project. The special conditions for a construction contract usually are addressed in Division 00 of the specifications (see Table 2.1).

The technical specifications provide the qualitative requirements for construction materials, equipment to be installed, and workmanship. They typically are organized using the MasterFormat shown in Table 2.1 that was developed by the Construction Specifications Institute (CSI). Each division of the specifications is further subdivided using a standard six-digit code developed by CSI. For example, Section 064023 refers to interior architectural millwork. Prior to 2004, the MasterFormat had fewer divisions, which are shown in Table 2.2. Some design firms may still be using this older structure in organizing their specifications.

There are four types of technical specifications used in construction contracts:

- descriptive specifications detail the required properties and quality of a material, assembly of materials, or products;
- performance specifications establish functional criteria for acceptable materials or products;
- proprietary specifications identify acceptable products by brand name, model number, or trade name; and
- reference standard specifications reference well-known standards such as the American Society for Testing and Materials (ASTM), American Concrete Institute (ACI), National Electrical Code (NEC), National Plumbing Code (NPC), and National Fire Protection Association (NFPA).

The types of specifications are not mutually exclusive. A single specification may contain descriptive and reference requirements.

**Table 2.1**  MasterFormat for technical specifications

| Division 00 | Procurement and contracting requirements | Division 26 | Electrical |
|---|---|---|---|
| | | Division 27 | Communications |
| Division 01 | General requirements | Division 28 | Electrical safety and security |
| Division 02 | Existing conditions | Division 31 | Earthwork |
| Division 03 | Concrete | Division 32 | Exterior improvements |
| Division 04 | Masonry | Division 33 | Utilities |
| Division 05 | Metals | Division 34 | Transportation |
| Division 06 | Wood, plastics, and composites | Division 35 | Waterway and marine construction |
| Division 07 | Thermal and moisture protection | Division 40 | Process interconnections |
| Division 08 | Openings | Division 41 | Material processing and handling equipment |
| Division 09 | Finishes | | |
| Division 10 | Specialties | Division 42 | Process heating, cooling, and drying equipment |
| Division 11 | Equipment | | |
| Division 12 | Furnishings | Division 43 | Process gas and liquid handling, purification, and storage equipment |
| Division 13 | Special construction | | |
| Division 14 | Conveying equipment | Division 44 | Pollution and waste control equipment |
| Division 21 | Fire suppression | Division 45 | Industry-specific manufacturing equipment |
| Division 22 | Plumbing | | |
| Division 23 | Heating, ventilating, and air conditioning | Division 46 | Water and wastewater |
| Division 25 | Integrated automation | Division 48 | Electrical power generation |

**Table 2.2**  Pre-2004 MasterFormat for technical specifications

| Division 00 | Supplemental conditions | Division 09 | Finishes |
|---|---|---|---|
| Division 01 | General requirements | Division 10 | Specialties |
| Division 02 | Site construction | Division 11 | Equipment |
| Division 03 | Concrete | Division 12 | Furnishings |
| Division 04 | Masonry | Division 13 | Special Systems |
| Division 05 | Metals | Division 14 | Hoisting Systems |
| Division 06 | Wood, plastics, and composites | Division 15 | Mechanical |
| Division 07 | Thermal and moisture protection | Division 16 | Electrical |
| Division 08 | Openings | | |

The drawings show the quantitative requirements for the project and how the various components go together to form the completed project. The drawings may be printed on paper or produced electronically. The initial set of drawings issued to the contractor is updated, as required, throughout the duration of the project to reflect modifications made by change orders. These updated construction drawings become the basis for the set of as-built (or record) drawings that the contractor provides to the owner at project close-out. The drawings for a typical construction contract are organized as shown below. Not all of these types of drawings will be used on all projects.

- architectural drawings;
- controls drawings;
- electrical drawings;
- equipment drawings;
- fire protection drawings;
- landscaping drawings;
- low voltage data systems drawings;
- mechanical drawings;
- plumbing drawings;
- shoring drawings;
- site or civil drawings;
- structural drawings; and
- room finish, light fixture, and plumbing fixture schedules.

Contracts awarded in the construction manager-at-risk and design-build project delivery methods typically do not have complete drawings and specifications. This is because the construction manager-at-risk contract is awarded early in the design process before many design decisions are finalized, and the design-build contract is awarded before design is started.

## Procurement methods

Owners who procure contracts use either a bid or a negotiated procedure. Public owners, such as government agencies, use public solicitation or procurement methods. These owners may require potential contractors to submit documentation of their qualifications for review before being allowed to submit a bid or proposal, or the owners may open the solicitations to all qualified contractors. The first method is known as prequalification of contractors, and only the most qualified contractors are invited to submit a bid or proposal. Private owners can use any method they like to select a contractor. Private owners often use contractors they have had good experience with in the past and may ask a select few or even only one contractor to submit a proposal.

## Bid method

Bid contracts generally are awarded solely on price. The owner defines the scope of the project, and the contractors submit lump sum bids, unit price bids, or a combination of both. The owner awards the contract to the contractor submitting the lowest total price for the project. Since actual quantities of work are not known for unit priced items, the contractor's unit prices are multiplied by the estimated quantities provided on the bid form by the owner and summed. The contractor submitting the lowest sum is selected for award. The steps in awarding a contract using a bid procedure are shown in Figure 2.2. The pre-bid conference usually is held in the designer's office or at the project site to resolve any contractors' questions relating to the project or the contract.

A lump sum contract may require a single price for the entire scope of work or require separate prices for individual portions of the scope of work. Some contracts may have additive or deductive items that must be priced during the bidding process. The owner selects which combination of additive and/or deductive alternates to award once the bids have been opened. Box 2.1 illustrates a bid form containing additive and deductive alternates. Some contracts may contain a combination of lump sum and unit price items as shown in Box 2.2.

## Negotiated method

Negotiated contracts are awarded based on any criteria the owner selects. Typical criteria include: cost (or fee in the case of a cost-plus contract), project duration, expertise of the project management team, plan for managing the project, contractor's safety record, contractor's existing work load, and contractor's experience with similar projects. Most negotiated contracts procured by public owners involve a two-step procedure. First, prospective contractors are prequalified after review of their prior work experiences and safety records. The first step or prequalification procedure is shown in Figure 2.3.

Then a short list of the most qualified contractors (generally three to five) are invited to submit proposals containing project specific information required by the owner. As a part of the evaluation procedure, owners may require the proposed project management teams to brief their plans for managing the project. This may include preparation of a project schedule and budget. The owner then selects the contractor submitting the best proposal and negotiates a contract price, and maybe a duration. The steps in awarding a contract using a negotiated procedure are shown in Figure 2.4. The pre-proposal conference is similar to the pre-bid conference used in a bid procedure. The major difference in a negotiated procedure is the opportunity for the owner to discuss the contractors' proposals, modify contract requirements, and clarify any issues before requesting best and final offers. The owner selects the contractor submitting the best value final proposal, which may not be the least cost.

Some owners use a more informal negotiating procedure, particularly if they have long-term relationships with their contractors. Such an owner may simply ask one or a few contractors to submit proposals. After reviewing the proposals, the owner would negotiate contract terms with the selected contractor.

## Contract pricing methods

There are several methods for pricing contracts used in the construction industry. The choice of which to use on a particular project is made by the owner after analyzing the risk associated with the project and deciding how much of the risk to assume and how much to impose on the contractor. Contractors want compensation for risk they assume.

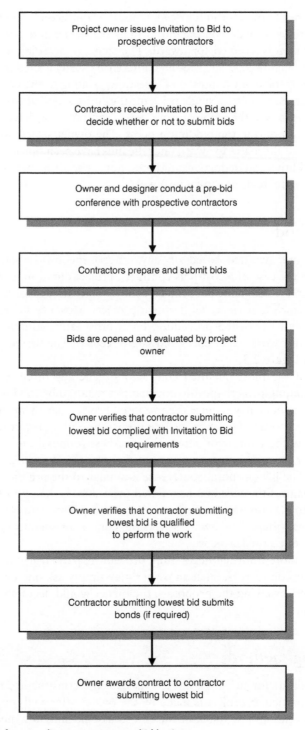

**Figure 2.2** Procedure for awarding a contract on a bid basis

# Box 2.1   Contract bid form containing alternates

**<u>BASE BID</u>**

Pursuant to and in compliance with the Advertisement for Bids and Instructions to Bidders, the undersigned hereby certifies having carefully examined the Contract Documents entitled

South Central High School,
prepared by Stellar Architects

and conditions affecting the work, and is familiar with the site; and having made the necessary examinations, hereby proposes to furnish all labor, materials, equipment, and services necessary to complete the work in strict accordance with the above named documents for the sum of

_____ Dollars ($_____)

which sum is hereby designated as the Base Bid.

**<u>ALTERNATES</u>**

The undersigned proposes to perform work called for in the following alternates, as described in Section 010300 and the drawings of the Contract Documents, for the following resulting additions to or deductions from the Base Bid.

Alternate #1:  Delete Selected Landscaping

_____ Dollars ($_____)

Alternate #2:  Delete Paved Parking Lot

_____ Dollars ($_____)

Alternate #3:  Add First Floor Upgrade

_____ Dollars ($_____)

Alternate #4:  Add Second Floor Upgrade

_____ Dollars ($_____)

Legal Name of Bidder: _____

By: _____
Signature/Title

Date: _____

## Box 2.2 Contract bid form containing lump sum and unit price bid items

<u>BID</u>

Pursuant to and in compliance with the Advertisement for Bids and Instructions to Bidders, the undersigned hereby certifies having carefully examined the Contract Documents entitled

Olympic Office Tower,
prepared by Cascade Designers, Inc.

and conditions affecting the work, and is familiar with the site; and having made the necessary examinations, hereby proposes to furnish all labor, materials, equipment, and services necessary to complete the work, less the drilled pier foundations, in strict accordance with the above named documents for the sum of

_____ Dollars ($_____)

The undersigned proposes to furnish all labor, materials, equipment, and services necessary to construct the drilled pier foundation for the following schedule of prices.  Exact quantities will be determined upon completion of the work.

| Item | Est. Quantity | Unit | Unit Price | Amount |
|---|---|---|---|---|
| 24-inch diameter drilled piers | 600 | linear feet | | |
| 36-inch diameter drilled piers | 800 | linear feet | | |
| 48-inch diameter drilled piers | 900 | linear feet | | |

Total for Unit Priced Items:

_____ Dollars ($_____)

Total Bid Price:

_____ Dollars ($_____)

Legal Name of Bidder: _____

By: _____
                              Signature/Title
Date: _____

## Lump sum

Lump sum (sometimes called *stipulated-sum* or *fixed-price*) contracts are used when the scope of work can be defined. The owner provides a set of drawings and specifications, and the contractor agrees to complete the project for a lump sum. Lump sum contracts also often are used for design-build projects where the owner specifies design criteria, and the contractor agrees to design and construct the project for a single price. While the exact scope of work is not defined in a design-

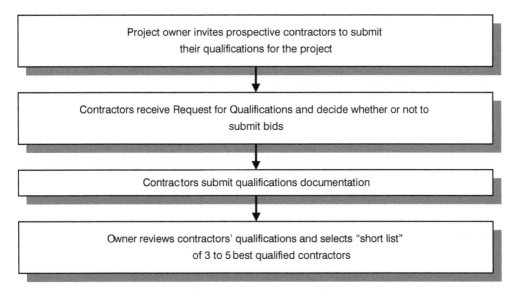

**Figure 2.3** Prequalification procedure for awarding a contract on a negotiated basis

build project, the contractor controls the design process and produces a design that can be built for the contract price. In a lump sum contract, the contractor is responsible for determining all the material, labor, equipment, and subcontract costs to establish the project cost. The initial contract price accepted by the owner at contract award may be modified during the life of the project by contract change orders, as we will see in Chapter 15.

## Unit price

Unit price contracts are used when the exact quantities of work are not known at the time the contract is executed. The designer provides an estimate of the quantity of each material required, and the contractor determines a unit price for each material. An example of a unit price bid schedule for a utility project is shown in Table 2.3. The cost data shown in italics would be entered on the form by the contractor and submitted to the owner as the bid for the project. The actual contract value for each item is not determined until the completion of the project. The actual quantities of work are measured throughout the completion of the project, and the cost is determined by multiplying the actual quantity by the unit price established by the contractor. Unit price contracts are used extensively on highway jobs or environmental clean-up jobs where exact quantities of work are difficult to define. Unit price and lump sum methods can both be used in the same contract. Sometimes the foundation portion of a building may be unit price, while the remainder of the building is lump sum.

## Cost plus

Cost-plus contracts are used when the scope of work cannot be defined. They are sometimes referred to as *cost-reimbursable contracts*. All specified contractor's project-related costs are

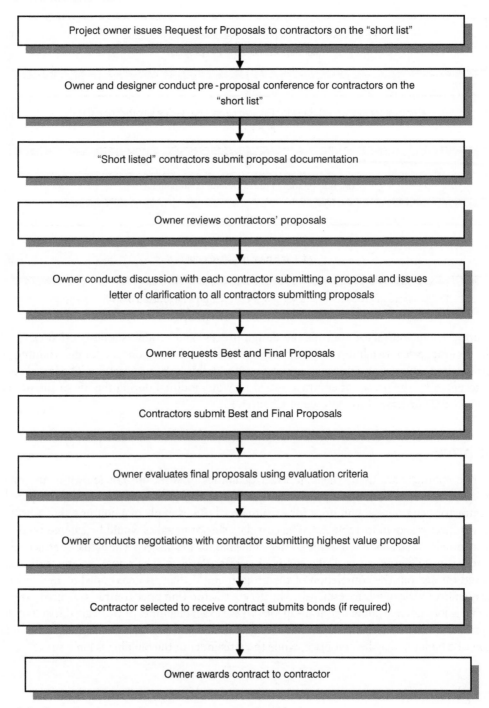

**Figure 2.4**    Procedure for awarding a contract on a negotiated basis

**Table 2.3**  Sample unit price bid form

| Work items | Unit | Estimated quantity | Unit price | Bid amount |
|---|---|---|---|---|
| Soil excavation | cubic yards | 12,000 | $6.00 | $72,000 |
| 12″ concrete pipe | linear feet | 1,000 | $18.00 | $18,000 |
| Crushed rock fill | cubic yards | 3,000 | $24.00 | $72,000 |
| Compacted fill | cubic yards | 9,000 | $15.00 | $135,000 |
| TOTAL | | | | $297,000 |

reimbursed by the owner, and a fee is added to cover profit and company overhead. Only those costs identified in the contract are reimbursable. The fee may be a fixed amount or a percentage of project costs or may have an incentive component. The fee in a cost-plus-fixed-fee contract is a fixed amount irrespective of the actual construction cost. The fee in a cost-plus-percentage-fee contract is a percentage of the actual project cost. The fee in a cost-plus-incentive-fee or cost-plus-award-fee contract is a variable based on the contractor's performance. The owner establishes a set of criteria that are used to determine the actual fee.

Cost-plus contracts typically are awarded based on an agreed fee structure, as project costs are not known until the project has been completed. Cost-plus contracts generally are used when the construction contract is awarded prior to the completion of the design. They also frequently are used in emergency conditions such as reconstruction after a major natural disaster, such as a hurricane, earthquake, or flood.

A special type of a cost-plus contract is known as a time-and-materials contract. In this type of contract, the owner and the contractor agree to a labor rate that includes the contractor's profit and overhead. Reimbursement to the contractor is made based on actual costs for materials and the agreed labor rate times the number of hours worked. Time-and-materials contracts generally are used only for small projects, maintenance and repair, or material testing.

## Cost plus with guaranteed maximum price

A guaranteed maximum price (GMP) contract is a type of cost-plus contract in which the contractor agrees to construct the project at or below a specified cost. Any cost exceeding the GMP would be borne by the contractor. Some of these contracts have a saving-sharing formula if the actual cost is less than the guaranteed maximum value. This is to provide an incentive to the contractor to control costs. For example, 20 percent of the savings might go to the contractor and 80 percent to the owner. In other GMP contracts, all the savings accrue to the owner.

## Standard contract forms

Contracts are either standard or specially prepared agreements. Most government agencies use standard formats for developing construction contract documents. Federal and state agencies typically have standardized general conditions and agreement language that is prescribed by government procurement regulations. Many local government agencies and private owners use contract formats developed the following professional organizations:

- ConsensusDocs (www.ConsensusDocs.org);
- American Institute of Architects (AIA) (www.aia.org/contractdocs/index.htm);
- Engineers Joint Council Document Committee (EJCDC) (www.ejcdc.org/); and
- Design Build Institute of America (DBIA) (www.dbia.org/resource-center/Pages/Contracts.aspx).

Contracts should not be signed until they have been subjected to a thorough legal review. This is to ensure that the documents are legally enforceable in the event of a disagreement and that there is a clear, legal description of each party's responsibilities. The advantage of using standard government or professional copyright contract forms is that they have been developed by individuals skilled at contract law and have been tested in and out of courts. Care should be exercised when modifying any of these standard forms. Most are available in electronic format and  are easy to tailor for specific projects. We will illustrate the use of selected ConsensusDocs contract forms throughout this text.

## Bonds

Bonds are used to protect the owner from some of the risks associated with construction. The concept of a bond is that a third party (called the *surety*) guarantees to the owner that the contractor will perform, as illustrated in Figure 2.5. When the contractor completes its obligations under the terms of the contract, the bond expires. Most public owners require bonds on projects exceeding a certain value, such as $250,000. Similar bonds also may be required by general contractors to guarantee the performance of subcontractors. This is discussed in Chapter 6.

There are four types of bonds typically used in construction:

- Bid bond: guarantees that the low bidder will submit performance and payment bonds and execute a contract to complete the project at the price bid. If the contractor does not, the bid bond becomes payable to the owner as compensation for damages sustained, which are the additional costs incurred by awarding the contract to the next lowest bidder. An example of a bid bond is shown in Box 2.3.
- Performance bond: guarantees that the contractor will complete the project in accordance with the contract plans and specifications. If the contractor fails to perform or defaults, the bonding company either must complete the project in accordance with contract requirements or compensate the owner for additional costs incurred by hiring another contractor to complete the project. An example performance bond is shown in Box 2.4.
- Labor and material payment bond: guarantees that the contractor will pay all suppliers, and subcontractors and that no liens will be placed on the completed project. If the contractor is unable or refuses to pay for work performed or materials used on the project, the surety will pay the claimants. An example labor and material payment bond is shown in Box 2.5.
- Maintenance bond (sometimes called a *warranty bond*): guarantees that the contractor will return to the completed project and repair or replace any defective or inferior materials or

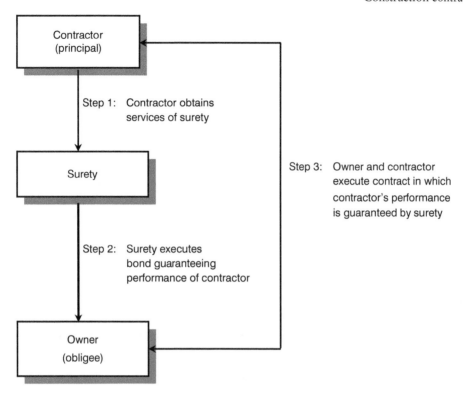

**Figure 2.5** Relationship between owner, surety, and contractor

workmanship during the specified warranty period. Maintenance bonds may be required by project owners for components with warranties that exceed the normal 1-year warranty, such as a 15-year warranty on a roof. An example maintenance bond is shown in Box 2.6.

Private owners may or may not require bonds on their projects. However, lending institutions often require bonds before financing private projects. Contractors purchase the bonds from a surety (also known as a bonding company) typically through a broker. Surety underwriters determine the amount of bonding that they will provide to an individual contractor after evaluating the character, capacity, and capital resources of the contractor. Capital refers to the financial strength of the contractor, capacity refers to the number and sizes of projects the contractor can handle, and character refers to the contractor's reputation for completing projects on time and satisfying owners. A sample bond rate schedule for a medium size general contractor with a good reputation and good financial condition is shown in Table 2.4. Rates will vary among contractors based on the underwriter's assessment and market conditions. Rates shown are for combined performance and labor and material payment bonds. Some sureties combine performance and labor and material payment provisions into a single bond. Bid bonds are provided at no cost to the contractor since there is little risk that the contractor will refuse to accept the contract, and the bonding company wants to sell the performance and labor and material payment bonds to the contractor.

## Box 2.3   Bid bond

**Bond No. BQ-157**

KNOW ALL BY THESE PRESENTS, that we, Northwest Construction Company of Cascade, Washington (herein called the Principal) as Principal and American Surety Company, a New York Corporation of New York, New York (herein called the Surety) as Surety are held and firmly bound unto the City of Seattle (herein called the Obligee) in the penal sum of nine hundred eighty-five thousand dollars ($985,000) for the payment of which the Principal and the Surety bind themselves, their heirs, executors, administrators, successors and assigns, jointly and severely, firmly by these presents.

THE CONDITION OF THIS OBLIGATION IS SUCH, that WHEREAS the Principal has submitted a bid to the Obligee on a contract for construction of the Sand Point Library.

NOW, THEREFORE, if the said contract be timely awarded to the Principal and if the Principal shall, within such time as may be specified, enter into the contract and give Bond with surety acceptable to the Obligee for performance of said contract, then this obligation shall be void; otherwise to remain in full force and effect.

Signed and sealed this fifth day of January, 2017.

*Harry P. Jones*
(Witness)

*Sam Peters*
Sam Peters, Vice President
Northwest Construction Company
(Principal)

*Samuel T. White*
(Witness)

*Terry L. Johnson*
Terry L. Johnson, Attorney-in-Fact
American Surety Company
(Surety)

Bid bonds usually are one-page documents that must be submitted with the bid documents. Some owners allow a certified cashier check to be submitted in lieu of a bond as bid security. Performance and payment bonds are not submitted until the owner indicates its intention to award the contract to a specific contractor. Bond coverage requirements specified by owners typically are 5 percent to 10 percent of the contract price for bid bonds and 50 percent to 100 percent for performance and payment bonds. Bonds impose no obligation on either the surety or the contractor beyond that contained in the construction contract.

# Box 2.4   Performance bond

**American Surety Company**
**126 William Street**
**New York, New York 10038**

Bond No. BP-157

KNOW ALL BY THESE PRESENTS, that we, Northwest Construction Company of Cascade, Washington (herein called the Principal) as Principal and American Surety Company, a New York Corporation of New York, New York (herein called the Surety) as Surety are held and firmly bound unto the University of Washington (herein called the Obligee) in the penal sum of sixty-one million, twenty-four thousand dollars ($61,024,000) for the payment of which the Principal and the Surety bind themselves, their heirs, executors, administrators, successors and assigns, jointly and severely, firmly by these presents.

THE CONDITION OF THIS OBLIGATION IS SUCH, that WHEREAS the Principal has entered into a written contract with said Obligee, dated March 24, 2015, for the NanoEngineering Building in accordance with the terms and conditions of said contract, which is hereby referred to and made a part hereof as if fully set forth herein.

NOW, THEREFORE, if the Principal shall promptly and faithfully perform said contract and reimburse the Obligee for all loss and damage sustained by reason of failure or default on the part of the Principal, then this obligation shall be void; otherwise to remain in full force and effect.

Signed and sealed this twenty-fourth day of March, 2015.

*Harry P. Jones*
(Witness)

*Sam Peters*
Sam Peters, Vice President
Northwest Construction Company
(Principal)

*Samuel T. White*
(Witness)

*Terry L. Johnson*
Terry L. Johnson, Attorney-in-Fact
American Surety Company
(Surety)

## Box 2.5    Labor and material payment bond

**American Surety Company**
**126 William Street**
**New York, New York 10038**

**Bond No. LM -157**

KNOW ALL BY THESE PRESENTS, that we, Northwest Construction Company of Cascade, Washington (herein called the Principal) as Principal and American Surety Company, a New York Corporation of New York, New York (herein called the Surety) as Surety are held and firmly bound unto the University of Washington (herein called the Obligee) in the penal sum of sixty-one million, twenty-four thousand dollars ($61,024,000) for the payment of which the Principal and the Surety bind themselves, their heirs, executors, administrators, successors and assigns, jointly and severely, firmly by these presents.

THE CONDITION OF THIS OBLIGATION IS SUCH, that WHEREAS the Principal has entered into a written contract with said Obligee, dated March 24, 2015, for the NanoEngineering Building in accordance with the terms and conditions of said contract, which is hereby referred to and made a part hereof as if fully set forth herein.

NOW, THEREFORE, if the Principal shall promptly and faithfully make payment to all Claimants for all labor and material used or reasonably required for use in the performance of the contract, then this obligation shall be void; otherwise to remain in full force and effect.

Signed and sealed this twenty-fourth day of March, 2015.

*Harry P. Jones*
(Witness)

*Sam Peters*
Sam Peters, Vice President
Northwest Construction Company
(Principal)

*Samuel T. White*
(Witness)

*Terry L. Johnson*
Terry L. Johnson, Attorney-in-Fact
American Surety Company
(Surety)

# Box 2.6    Maintenance bond

**American Surety Company**
**126 William Street**
**New York, New York 10038**

**Bond No. MT-145**

KNOW ALL BY THESE PRESENTS, that we, Northwest Construction Company of Seattle, Washington (herein called the Principal) as Principal and American Surety Company, a New York Corporation of New York, New York (herein called the Surety) as Surety are held and firmly bound unto the Emerald City Development Corporation (herein called the Obligee) in the penal sum of eight hundred fifty thousand dollars ($850,000) for the payment of which the Principal and the Surety bind themselves, their heirs, executors, administrators, successors and assigns, jointly and severely, firmly by these presents.

THE CONDITION OF THIS OBLIGATION IS SUCH, that WHEREAS the Principal has entered into a written contract with said Obligee, dated March 18, 2016, for the Cascade Medical Office Building that provides that the Principal will furnish a bond conditioned to guarantee the roof of said facility for a period of 10 years against all defects in workmanship and materials which may become apparent during said period, said contract, which is hereby referred to and made a part hereof as if fully set forth herein, and

WHEREAS said contract was completed on January 4, 2017.

NOW, THEREFORE, if the Principal shall promptly and faithfully make good, repair, and replace at its own expense any defects discovered prior to January 4, 2028  in work done or material furnished under said contract relating to the roof, then this obligation shall be void; otherwise to remain in full force and effect.

Signed and sealed this fourth day of January, 2017.

| | |
|---|---|
| *Harry P. Jones* | *Sam Peters* |
| (Witness) | Sam Peters, Vice President |
| | Northwest Construction Company |
| | (Principal) |
| | |
| *Samuel T. White* | *Terry L. Johnson* |
| (Witness) | Terry L. Johnson, Attorney-in-Fact |
| | American Surety Company |
| | (Surety) |

**Table 2.4**   Sample bond rate schedule

| Contract price | Premium cost per $1,000 of contract price for the first 12 months of a project (Additional Time Would Add Additional Cost) |
|---|---|
| $1 to $100,000 | $30 |
| Next $400,000 | $20 |
| Next $1,000,000 | $15 |
| Next $2,000,000 | $10 |
| Next $2,000,000 | $7 |
| Next $2,000,000 | $6 |
| Over $7,500,000 | $5 |

# Insurance

Insurance is purchased by the contractor to cover some of the risks associated with the project. An insurance policy is a two-party contract under which the insurer (insurance company) promises, for a consideration (fee or premium), to assume the financial responsibility for a specified loss or liability for a specified time. The amount of risk transferred is limited to the amounts stipulated in the individual insurance policies. Minimum coverage requirements generally are specified in the contract documents, and deductibles may be used to reduce insurance premiums if the contractor is willing to assume a portion of the risk. The following eight types of insurance policies typically are purchased by the contractor:

- Builder's risk insurance: protects the contractor in the event the project is damaged or destroyed while under construction. Builder's risk insurance may be all-risk or limited to named perils such as fire, wind, hail, and explosion. All-risk insurance covers all damage unless there is a risk exclusion such as earthquake. Builder's risk insurance generally includes coverage of materials stored on site and may include coverage of materials in transit and stored off site. Builder's risk policies typically include a standard exclusion for loss or damage resulting from faulty or defective workmanship or material. Damage to contractor's equipment stored on the construction site typically is not covered by builder's risk insurance. Builder's risk insurance is typically purchased with coverage extending to the general contractor, the project owner, and all subcontractors working on the project. The subcontractors are usually listed as additional insured on the builder's risk insurance policy.
- General liability insurance: protects the contractor against financial losses that may result from injuries or property losses sustained by third parties as a consequence of the contractor's activities.
- Property damage insurance: protects the contractor against financial loss due to damage to the contractor's property, such as an office building, warehouse, or maintenance facility.
- Equipment floater insurance: protects the contractor against financial loss due to physical damage to equipment from named perils or all risks and theft. Coverage is for owned, leased, and rented equipment not operated on streets and highways.

- Automobile insurance: protects the contractor against claims from another party for bodily injury or property damage caused by contractor owned, leased, or rented automobiles and equipment operated over the highway. Coverage may include damages to the automobiles and equipment.
- Umbrella liability insurance: provides coverage against liability claims exceeding that covered by standard general liability or automobile insurance. For example, a contractor may have a general liability insurance policy covering up to $2 million per occurrence and an umbrella policy covering up to $50 million per occurrence.
- Workers' compensation insurance: protects the contractor from a claim due to injury or death of an employee on the project site. Workers' compensation insurance is no-fault insurance in which the employer cannot deny a claim by an insured employee and the employee cannot sue the employer for injuries sustained on the job.
- Errors and omissions insurance: protects the contractor against claims due to the professional negligence of design professionals that the contractor may employ.

Some owners purchase the builder's risk insurance, while others require the contractor to purchase it. Coverage expires upon acceptance of the completed project by the owner. Workers' compensation insurance is either purchased from a commercial insurance company or a state fund. Some state governments require employers to purchase worker's compensation insurance from a monopolistic state fund, unless they choose to be self-insuring. To be self-insuring, the contractor typically must meet certain size and financial requirements.

## Impact on project management

The contract documents have a significant impact on the responsibilities of the project manager. Specific requirements are contained in the general conditions and special conditions of the contract. Unit price and cost-plus contracts have more owner involvement than do lump sum contracts. The actual quantities of work are jointly determined by the owner and the project manager in a unit price contract. Project expenditures often require owner approval in cost-plus contracts, and invoices must be collected on the project site and submitted to the owner for reimbursement. A comparison of lump sum and cost-plus contracts from a contractor's perspective is shown in Table 2.5.

**Table 2.5** Comparison of lump sum and cost-plus contracts

| Lump sum contract | Cost-plus contract |
| --- | --- |
| Riskiest to contractor | Least risk for contractor |
| Sometimes adversarial | Team atmosphere |
| Little owner input to project management | May be significant owner involvement in project management |
| Owner controls funding on disputed work | All direct project cost reimbursed |
| Savings resulting from innovation go to contractor | Savings resulting from innovation may be shared between owner and contractor |

## Selecting a project

Now that we have discussed basic project management concepts and organizations and construction contracts, we will examine selection and acquisition of a project to manage. In this section, we will discuss analyzing the risks associated with a project, techniques for convincing an owner to select a general contractor to manage his or her project, and strategies for bidding or negotiating for the project. In many construction firms, the selection of projects to pursue is made by senior executives, not individual project managers. In other firms, the project managers recommend projects for senior executive approval. At the end of the chapter, you will be introduced to an example project that we will use throughout the remainder of the book to illustrate specific concepts and procedures.

# Risk analysis

Construction is a risky business, as evidenced by the high number of construction firm failures each year. To minimize the potential for financial difficulty, a contractor should analyze each potential project to determine the risks involved and whether the potential rewards justify acceptance of the risk exposure. The basic steps in performing a risk analysis are:

- identify the sources of risk,
- identify the range of potential risk events,
- assess the potential impacts of risk events on the project,
- identify alternative responses to mitigate the potential impacts of risk events,
- identify the consequences of the alternative responses, and
- select risk management strategies including the allocation of risk.

Some of the sources of risk on a project may involve such things as material cost inflation, owner's inability to finance the project, limited availability of skilled craftspeople or subcontractors, bankruptcy of subcontractors, incomplete design documents, project size, project location, and project constructability and complexity. The contractor needs to forecast the likelihood of such risks, the range of possibilities, and the impact of each on the contractor's ability to complete the project profitably.

The primary risk responses are:

- avoiding the risk by walking away from the project,
- mitigating the risk by sharing it with a joint-venture partner or hiring a subcontractor,
- increasing the fee requested,
- transferring the risk by purchasing insurance, or
- accepting the risk.

Selection of risk management strategies involves selection of the appropriate response to each of the identified risks. Insurance coverage generally is limited to protection against financial loss due to damage to the project under construction (builder's risk); injury of workers on the project site (workers' compensation); injury or property loss to a third party (liability); and damage to equipment. Most project owners require that the general contractor name the owner as an "additional insured" on the contractor's general liability and errors and omissions insurance policies to protect the owner against claims made due to actions taken by the contractor's employees.

Internal risks must be identified also and appropriate management strategies selected. The three most common internal risks are unrealistic cost estimates; unrealistic construction schedules; and ineffective project management including cost and schedule control, material management, and subcontractor coordination. Contractors must adopt strategies to minimize the potential of these problems occurring. Often the basic issue to be addressed is the selection of qualified people to manage the project, particularly the project manager and the superintendent.

The output of a risk analysis is a decision whether to pursue a project, the amount of contingency to include in the bid or cost proposal, the amount of fee to include in the bid or cost proposal, whether to joint-venture with another firm, the portions of work to subcontract, and the type and amount of insurance to purchase.

## Marketing

Marketing is acquiring and retaining customers. This is an essential project management skill, because the most effective marketing resource is a satisfied customer. In most projects, the customer is the owner, and customer satisfaction is measured in terms of cost of construction, time to complete, and the resulting quality of the completed project. These project characteristics are the responsibility of the project manager, as discussed in Chapter 1.

The initial marketing task is to determine which segment of the overall market the contractor wants for its customers. Some contractors may want to bid for public projects such as schools, government office buildings, or highways. Others may be interested only in privately funded projects such as shopping centers, private office buildings, hotels, or condominiums. Once the market segment has been selected, the contractor needs to target its marketing activities toward those prospective customers. Then, the contractor organizes its services to cater to the specific market segment selected. This requires an understanding regarding how the targeted owners select contractors for their projects and what project characteristics they value. For example, some private owners are more interested in high-quality, short duration projects than they are in least cost. The contractor should emphasize these characteristics in its marketing initiatives. For example, a contractor wanting to enter the hotel construction market will want to ensure that his or her project managers understand the prospective customers' requirements and are skilled at preparing proposals and making presentations for that specific market and owner.

Everything that occurs on a project should be a part of the contractor's marketing strategy. The project sign, the organization and cleanliness of the project site, the treatment of subcontractors and suppliers, the condition of its equipment, and the accident rate all shape the public's and prospective owners' opinions regarding the contractor. Public relations are also a part of marketing. Participation in professional organizations and public service projects establishes a network of contacts for the project manager and significantly aids in marketing. Networking also results in new insights on marketing strategies and lessons learned.

A project manager and superintendent do not have a project to manage until one has been obtained. While the contractor's business development staff may identify potential clients, it usually is up to the project manager and superintendent to obtain the project. This may occur by preparing a bid for the project, or it may be by preparing a quality proposal and making a winning presentation to the potential customer. Good marketing and presentation skills are just as important as good technical skills to be a successful project manager. Once a project has been acquired, it is the responsibility of the project manager and superintendent to complete the project to the owner's requirements. This is to ensure the construction firm is considered for future projects built by the owner.

## Contact development

On a negotiated project, the construction contract can be executed early. Either a budget or an allowance may be used as a placeholder for the GMP until it is finalized, or the contractor may receive a letter of intent or a preconstruction agreement with the construction contract executed later after the design is nearing completion and the GMP finalized. Even though a proven copyrighted contract is used, many contractors, architects, and owners have standard revisions they make to the original wording. Beyond these changes, much of the contract language and execution is a process of filling in the blanks. The following are additional specific drafting guidelines for the contract agreement:

- Names and addresses should be shown on page one.
- Any previously issued letters of intent, letters of authorization, or preconstruction agreements may also be referred to and attached as exhibits.
- The contract defines the agreement. It is the most important document on the project and is to be all-encompassing. All relevant information and documents must be tied into the contract by reference and attached as exhibits. All exhibits are to be numbered or lettered, dated, and attached.
- Work should commence upon receipt of four items: (1) permits, (2) verification of financing, (3) notice to proceed, and (4) signed and executed contract. Specific dates that are out of the contractor's control should not be used.
- Required completion is established as calendar or work days after commencement. The project manager should here also avoid using a specific date that may be out of his or her control.
- The fee should either be a specific amount or a percentage of the actual cost. If there is an incentive component, the criteria for determining the fee should be explained. All parties should be clear about the fee amount.
- The general contractor may want to insert the project manager as a designated reimbursable cost, particularly if the project manager is dedicated only to this project.
- General contractors often negotiate an allowance for hand tools to avoid the necessity to track tool costs. If an allowance is to be used, its value should be stated.

The contract agreement is the most important construction document. All other documents on the project should be linked to it. The agreement should be self-sustaining and stand on its own. It should make reference to all relevant supporting documents, and they should be attached as exhibits. Some of the documents that could be linked to the contract agreement are:

- construction schedule,
- specifications,
- project manual,

- drawings,
- document exhibit that lists all of the relevant documents,
- geotechnical report,
- bid addenda including pre-bid meeting notes,
- approved submittals (submittals are discussed in Chapter 7),
- the request for proposal,
- letter of intent to award a contract from the owner to the contractor,
- preconstruction contract between the owner and the contractor (discussed in Chapter 5),
- the contractor's prescribed bid form,
- pre-bid site inspection certification form, and
- either a summary of the contractor's estimate or the entire detailed estimate.

If these documents are to be considered as contract documents, they must be referenced in the contract agreement by title and date. In some contracts, each party will sign or initial each exhibit similar to the execution of the contract itself.

When allowed by the owner, the project manager should participate in developing the construction contract agreement. This section discusses several basic principles that should be followed when preparing owner/contractor contracts, regardless of contract format.

In many cases, the contract itself is included or referred to in the request for proposal (RFP). This is a good practice from the owner's position. The party issuing the RFP may even go so far as to have the bidders acknowledge the contract format to be used and agree to accept its terms and conditions without any revisions. Wording similar to the following often is used in RFPs:

*Submittal of a proposal by a general contractor shall serve as evidence and acceptance of the contract format to be used and enclosed herein without any modifications, except as proposed by the contractor, for acceptance by the owner, and noted and attached to the contractor's bid proposal.*

In these cases, the project manager has no choice regarding the contract and must make a decision whether or not to bid the project.

When drafting or executing a contract, project managers should do as follows:

- Request two blank copies of the contract or an electronic version, if the format is specified by the owner.
- If no format is specified, select the standard contract document developed by the association that best represents his or her firm's business and the specific project.
- Avoid the use of double-sided documents. Sometimes the back sides of pages do not get copied and, without proper initialing of each page, as discussed below, cases have been made that the reverse side was not read.
- Request two original copies of the contract from the association.
- Make a photocopy of one of the originals and begin drafting the contract.
- Route drafts through the office for input from superiors and peers.
- Revise language based on company input and submit to the owner for review.
- Revise language, if required, based on the owner's input, and prepare two original documents. All exhibits that are referred to in the body of the contract must be attached.
- Initial each page of both originals. Pass the contract on to the officer-in-charge and ask for his or her initials on all pages as well as a signature on each of the originals in the appropriate blank.

- Initials and signatures should be original and not stamped, and they should be done with blue ink. Original black ink is sometimes difficult to discern from copies due to the high quality copy machines that are available. Pencil should not be used because it can be erased easily.
- Deliver both originals to the owner or architect, whichever is called out as the recipient and reviewer. Request that the owner also initial every page and return one executed contract back for the project manager's files. In this way, both the owner and the project manager will have an original copy.
- If the owner wishes to make a minor revision, he or she can simply mark out or add in text and initial the change on both originals. Then both originals will need to be returned to the project manager for counter-initial on the change. This allows the contract to be executed and avoids redrafting due to minor changes. If a major change is necessary, redrafting may be the only choice.

## Example project

The project selected as the example project for this text is a six-floor addition to an existing research building on the University of Washington campus in Seattle. The 78,000 gross square foot building includes an enlarged basement that extends to the north under an exterior plaza. An artist's rendering of the completed building is shown in Figure 2.6. The basic structure is cast-in-place concrete. The designer of the NanoEngineering Building was Zimmer Gunsul Frasca (ZGF), and the construction contractor was Hoffman Construction Company of Washington. The project delivery method selected for this project was construction manager-at-risk, and the contract was

**Figure 2.6**  Artist's rendering of completed case study project (image by ZGF architects, courtesy of University of Washington)

awarded early in the design process to enable the construction manager to perform preconstruction services while the design firm completed the design. The contract was awarded using a negotiated process and provides for a lump sum amount for preconstruction services, a lump sum for project general conditions, and a cost-plus fixed fee with GMP for construction. Because we are using fictitious contract documents in the book, we have chosen to use a fictitious construction company, Northwest Construction Company, so as not to imply that any of the documents that we developed were actually used on the project. A ConsensusDocs 500 contract agreement for the NanoEngineering Building project is available on the companion website for this book.

## Summary

A construction contract describes the responsibilities of the owner and the contractor and the terms and conditions of their relationship. A thorough understanding of all contractual requirements is essential, if a project manager expects to complete the project successfully.

There are five basic documents that comprise most construction contracts. They are the agreement, the general conditions, the special conditions, the technical specifications, and the contract drawings. The agreement describes the project to be constructed, the pricing method to be used, the time allowed for construction, any liquidated damages, and the names of the parties to the contract: the owner and the contractor. The general conditions of the contract provide a set of operating procedures that a specific owner typically uses on all projects. The special conditions of the contract contain any unique requirements for the project. Some construction contracts do not contain any special conditions. The technical specifications provide the qualitative requirements for construction materials, equipment to be installed, and workmanship. The contract drawings show the quantitative requirements for the project and how the various components go together to form the completed project.

There are several pricing methods that typically are used on construction contracts. Lump sum contracts are awarded on the basis of a single lump sum estimate for a specified scope of work. Unit price contracts are used when the exact quantities of work cannot be defined. The designer estimates the quantities of work, and the contractor submits unit prices for each work item. The actual quantities required are multiplied by the unit prices to determine the contract price. Cost-plus contracts are used when the scope of work cannot be defined. All the contractor's project-related costs are reimbursed by the owner, and a fee is paid to cover profit and contractor general overhead. A GMP contract is a cost-plus contract in which the contractor agrees not to exceed a specified cost.

Owners select contractors by one of two methods. In a bid procedure, the contractor is selected solely on the cost data submitted on the owner's bid form. In a negotiated procedure, the contractor is selected based on any criteria the owner wishes to establish.

Contracts are either standard or specifically prepared documents. Standard contracts generally are preferred because they have been legally tested in and out of courts. Such documents have been developed by several professional organizations.

Construction is a risky activity, and bonds and insurance are used by owners and contractors to cover some of the risks. Owners often require bid bonds to guarantee that the low bidder will accept the contract, performance bonds to guarantee that the contractor will complete the project, and labor and materials payment bonds to guarantee that there will be no liens placed on the project. Contractors purchase insurance contracts to protect against financial loss due to damage to the uncompleted project (builder's risk), claims from third parties (general liability), damage to their property (property damage), damage to or loss of equipment (equipment floater), and claims resulting from injury or death of an employee on the project (worker's compensation).

## Review questions

1. Why is it essential that a project manager fully understand the requirements and procedures specified in the contract documents?
2. What types of information would you find in each of the following contract documents?
    a. Agreement
    b. General conditions
    c. Special conditions
    d. Technical specifications
    e. Contract drawings
3. What is the difference between a contract addenda and a contract change order?
4. Why might an owner decide to include a special conditions section in a construction contract?
5. Name and describe the four types of technical specifications.
6. What type of construction materials are covered in division 3 of the technical specifications of a construction contract?
7. What type of construction materials are covered in division 9 of the technical specifications of a construction contract?
8. What is the difference between a lump sum and a unit price contract?
9. Under what conditions might an owner decide to use a unit price contract?
10. What is the difference between a lump sum and a cost-plus construction contract?
11. What is the difference between the bid and the negotiated methods of contract procurement?
12. On what basis is the successful contractor selected in a bid method of contract procurement?
13. On what basis is the successful contractor selected in a negotiated method of contract procurement?
14. What does contractor prequalification mean in the context of a construction contract procurement strategy?
15. What are the four types of bonds used in construction? Why do owners require them?
16. What is the difference between builder's risk and general liability insurance?
17. What is the difference between named-peril and all-risk builder's risk insurance?
18. What contractor's risk is covered by worker's compensation insurance?
19. What are five potential risks that a contractor might face on a construction project?
20. What are four alternative risk response strategies that a contractor might consider in developing a risk management plan?

## Exercises

1. What requirements might be included in the special conditions of a contract for the construction of a major addition to an operational hospital in a major metropolitan area? What impacts would these requirements have on the scheduled completion and cost of the project?

2. What information might an owner require contractors to submit for the owner's review during the prequalification process, if the owner wants to identify the best-qualified contractors for negotiating a design-build contract for the construction of a major manufacturing plant?

3. What criteria might an owner use when negotiating a lump sum contract for the construction of a shopping complex that she desires to be completed prior to the start of the Christmas shopping season?

4. Calculate a contractor's cost for performance and labor and material payment bonds using the rate schedule shown in Table 2.4 for each of the following:

   a. A construction project has an estimated cost of $750,000. What is the resultant bond fee as a percentage of the estimated construction cost?

   b. A construction project has an estimated cost of $25 million. What is the resultant bond fee as a percentage of the estimated construction cost?

5. What risks does a contractor face when bidding on a construction project located in a city in which they have never had a previous project?

6. What criteria should a contractor use in deciding for which to submit bids?

# 3

# Cost estimating

## Introduction

As discussed in Chapter 1, cost is one of the most critical project attributes that must be controlled by the project manager and superintendent. Project costs are estimated to develop a project budget within which the project team must build the project. All project costs are estimated in preparing bids for lump sum or unit price contracts and negotiating the guaranteed maximum price on cost-plus contracts. Non-reimbursable costs are estimated in developing fee proposals for cost-plus contracts.

Cost estimating is the process of collecting, analyzing, and summarizing data in order to prepare an educated projection of the anticipated cost of a project. Project cost estimates may be prepared either by the project manager or by the estimating department of the construction firm. Even if the estimate is prepared by the estimating department, the project manager must understand how the estimate was prepared, because he or she along with the superintendent must build the project within the budget developed from the estimate. This project budget becomes the basis for the cost control system as will be discussed in Chapter 12.

Our goal in this chapter is not to reproduce all of the information that is available in other publications but to highlight some of the major issues in developing cost estimates. There are many good texts and references available on estimating. Many contractors use estimating software to develop their estimates. There are several on the market such as Timberline Precision, $MC^2$, WinEstimator, BID2WIN, HCSS, and ProEst, but many contractors use their own historical in-house databases with customized Excel spreadsheets.

Good cost estimating skills are essential if one is to be an effective project manager. The quantity take-off procedure for a few major building systems will be briefly discussed in this chapter. For discussion of other building systems, readers should refer to dedicated estimating texts such as *Construction Cost Estimating, Process and Practices*. If a project manager can calculate the quantity of concrete in a spot footing or count the quantity of hollow core wood doors, then he or she has the skills needed to measure almost any material quantity. The same rules of measuring, counting, and extending apply to most systems.

It is assumed that the reader has experience in construction means and methods. An understanding of the differences between basic building systems, such as concrete foundations from wood framing, is essential. It is also assumed that document reading is a tool you have already acquired and are familiar with terms and abbreviations such as SF (square foot) and CY (cubic yard).

There is no one exact estimate for any project. There are many correct estimates, although some are more correct than others. Adjustments in pricing, subcontractor and labor strategy, overhead structures, and fee calculations are individual contractor decisions that will determine "the estimate" for those conditions at that time. The process of developing an estimate is illustrated in Figure 3.1 depicted as an estimating triangle. The first step of gathering information is the foundation for the process. As the estimator proceeds through the process, information continues to be analyzed and summarized and refined, until eventually there is only one figure left: the final estimate or bid.

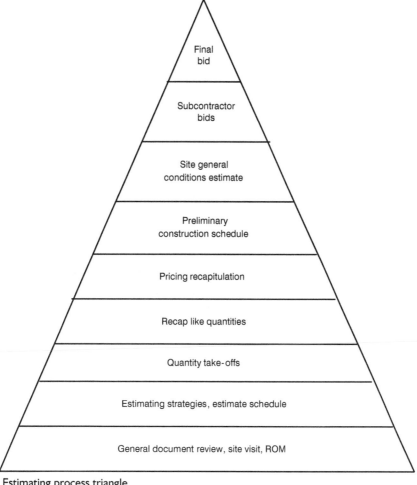

**Figure 3.1**   Estimating process triangle

**Table 3.1**   Estimate accuracies

| Type of estimate | Document development | Accuracy |
|---|---|---|
| Budget or rough order of magnitude (ROM) | Schematics, design development, 30% construction documents | 10–20% |
| Guaranteed maximum price for cost-plus projects | 60%–90% construction documents | 5% |
| Lump sum bid | 100% plans and specifications | 2% |

## Types of cost estimates

There are several different types of cost estimates. *Conceptual cost estimates* are developed using incomplete project documentation, while *detailed cost estimates* are prepared using complete drawings and specifications. *Semi-detailed cost estimates* are used for guaranteed maximum price contracts and have elements of both conceptual and detailed estimates, as discussed further below. The typical accuracy for each type of estimate is shown in Table 3.1. A type of conceptual cost estimate called a rough-order-of-magnitude (ROM) estimate is developed from the first document overview to establish a preliminary project budget and determine whether the contractor intends to pursue the project.

All estimates have the following major elements or categories, some of which require a more detailed effort by the estimator than others:

- direct labor,
- direct material,
- subcontractors and major material suppliers,
- jobsite administration or general conditions, including equipment rental, and
- percentage markups including fee, excise tax, contingency, and insurance.

## Risk analysis

The greatest risk in developing a cost estimate is estimating the productivity of the craft workers. This is where an experienced estimator has a great advantage. Other risks involve failure to include some element of work or double-counting another element of work. To minimize the potential for making errors when developing an estimate, the project manager/estimator should:

- rely on good estimating practices and procedures;
- choose good in-house project management and supervision teams not only to manage the project but also to assist with the estimating;
- choose qualified subcontractors and suppliers;
- plan to build the project in less time than specified in the contract to save jobsite overhead expenses; and
- be selective on which projects are chosen to bid. Familiarity with the owner, designer, and building location and type is important. Selection of projects may be made by senior managers in the construction firm, or sometimes they may ask for the project manager's recommendations.

# Estimating strategies

When possible, the project manager and superintendent should be responsible for developing the estimate or, at a minimum, work as integral members of the estimating team. Their individual inputs regarding constructability and their personal commitments to the estimating product are essential to ensure not only the success of the estimate but the ultimate success of the project. One of the first assignments for the project manager and the estimating team is to develop a responsibility list and to schedule the estimate. The estimating process should be scheduled for each project. Today's date is known, most probably a pre-bid or pre-proposal conference is scheduled, and the date the bid or proposal is due is also scheduled. With these milestones established, a short bar-chart schedule should be developed that shows each step and assigns due dates to each estimating task. Familiarity with the steps or building blocks shown in Figure 3.1 is essential in developing the schedule. Each team member is relying on the others to do their jobs efficiently and accurately. Similar to a construction schedule, if one of the individuals falls behind on any one activity, the completion date may be in jeopardy, or more commonly, the quality of the finished estimate will be affected unless other resources are applied. An estimate schedule for the NanoEngineering Building is shown in Figure 3.2. Many factors will affect the estimating strategy for each project. These include:

* project location,
* complexity of the project,
* type of contract,
* familiarity with the architect and engineers,
* workload of the construction company,
* season of the year,
* familiarity with the owner or client,
* adequacy of contract documents, and
* availability of estimating resources.

The estimating strategy or approach is different with each of the three main types of estimates; maybe even the type of estimator best qualified might vary. The level of detail will be different with each of the estimate types as well. Some of these approaches, strategies, and differences are reflected in Table 3.2. As indicated earlier, the accuracy of an estimate is directly proportional to the accuracy of the documents. The time spent on preparing an estimate also follows those same

**Table 3.2**  Estimating strategies

| Types of cost | Budget estimate | Semi-detailed or GMP estimate | Detailed or lump sum estimate |
|---|---|---|---|
| Direct Work: | | | |
| Labor and material: | Assembly pricing | Detailed QTO Competitive UPs | Detailed QTO Competitive UPs |
| Equipment rental: | with assemblies | with assemblies | Competitive pricing |
| Subcontractors: | In-house plugs | OM plugs or 1–3 quotes | Competitive sub quotes |
| General conditions: | Percentage add-on | Detailed estimate | Detailed estimate |

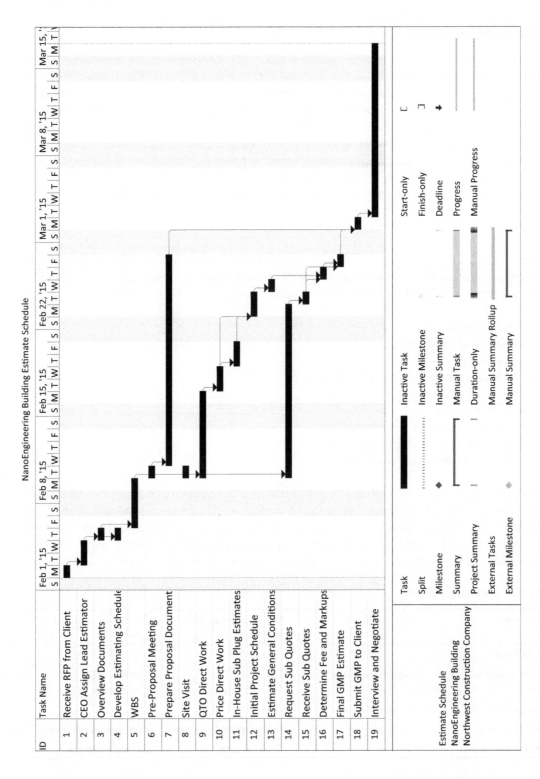

**Figure 3.2** Estimate schedule

lines. A fully detailed set of drawings is sufficient to prepare a lump sum estimate, where every item of work can be accurately measured and matched with unit prices or competitive subcontractor bids. A conceptual set of drawings can be estimated quite quickly using SF of floor unit prices, assembly prices, subcontractor plugs or budgets, and percentage add-ons for general conditions to produce a rough order of magnitude (ROM) budget. Subcontractor "plugs" are rough placeholder estimates generated by the general contractor based upon past similar projects. These plugs will be replaced once competitive bids are received from subcontractors. A guaranteed maximum price (GMP) estimate falls somewhere between. Detailed quantities and unit prices are utilized where possible, such as on structure; a mix of subcontractor quotes and budgets are added; and items not yet designed, such as architectural finishes, may still require a cost per SF of floor estimate.

## Work breakdown

The work breakdown structure (WBS) for the project is an early compilation of the significant work items that will have associated cost or schedule considerations. This includes areas of work such as foundations, utilities, drywall, floor covering, plumbing, and the like. Before any detailed estimating is performed, the estimator should have a general idea of the work that will be included on the WBS. The first step is to perform a document overview. The goal is to slowly leaf through the drawings and specifications and develop a good understanding of the type of project and the systems that are included. The estimator should not start quantifying or pricing any work items until this overview is complete.

One of the tools utilized to assist with this process is the development of a project item list. Worksheet 3.1 is an abbreviated example for a typical project. An expanded WBS is included on the companion website. This list is a good reference to use throughout the estimating process as well as a final checklist to review again prior to finalizing the estimate. This item list is not to be considered a final WBS, just a good first step.

At this stage, the estimator needs to know which categories of work will be estimated in detail and will be subsequently self-performed by the contractor's own forces and which will be supplied and installed by subcontractors. The project item list has boxes to check for this purpose. The types of work general contractors may perform include:

- hand excavation and backfill,
- concrete formwork,
- concrete reinforcement placement,
- concrete placement,
- concrete slab finishing,
- structural steel erection,
- rough carpentry,
- finish carpentry, including doors, specialties, and accessories,
- wood siding, and
- window placement.

Decisions regarding which scopes of work to self-perform and which to subcontract are based on several criteria.

## Worksheet 3.1    Project item list

**PROJECT ITEM LIST**

Project:    *Middle School Lump Sum Bid*        Date:  *2/6/2015*

Estimator:    *Ernie Sanchez*

| | | Provider: Subcontractor | | General Contractor / | |
|---|---|---|---|---|---|
| Item / CSI Div. | Cost Item Description | Material | Labor | Material | Labor |
| 1 | 2 | Demolition | | | X | X |
| 2 | 3 | Structural Excavation | | | X | X |
| 3 | 3 | CIP Concrete Formwork | X | X | | |
| 4 | 3 | Rebar | X | | | X |
| 5 | 3 | Embeds | X | X | | |
| 6 | 3 | Place Concrete | X | X | | |
| 7 | 3 | Concrete Pump | | | X | |
| 8 | 3 | Finish Flatwork | | | X | |
| 9 | 3 | PT Cables | | | X | X |
| 10 | 4 | CMU | | | X | X |
| 24 | 9 | ACT | | | X | X |
| 25 | 9 | Carpet | | | X | X |
| 26 | 9 | Ceramic Tile | | | X | X |
| 34 | 14 | Elevators | | | X | X |
| 35 | 21 | Fire Protection | | | X | X |
| 36 | 22 | Plumbing | | | X | X |
| 43 | 32 | Pavement Striping | | | X | X |
| 44 | 32 | Curbs and Sidewalks | | | X | X |
| 45 | 32 | Landscape and Irrigation | | | X | X |
| 46 | 33 | Site Utilities | | | X | X |

- Labor: Subcontractors must be used if the trades needed are not employed by the general contractor. Although shifting more work to subcontractors is a means of risk mitigation, there are also additional risks when subcontracting as discussed throughout this text.
- Specialization: If a subcontractor specializes in finishing concrete, it may have the most efficient and least expensive way to perform the work.

- Quality: If this is an area the subcontractor specializes in, it often can perform it better. Also, if there are problems with quality, a subcontractor is required to repair the work without increase in cost to the general contractor. However, if the general contractor is required to repair self-performed work, the general contractor's cost will increase, and the profit margin may correspondingly decrease.
- Price: If a subcontractor can perform the work on a fixed price contract for less than what the general contractor has estimated, with no guarantee of price, then the work may be awarded to the subcontractor.
- Work load: If the general contractor's labor forces are tied up on other work, some normally self-performed items may be awarded to subcontractors. Conversely, if the general contractor is low on work and wants to keep its labor force together, the contractor may choose to self-perform more work items.
- Schedule: Many general contractors will argue that they can control and ensure project schedule (and quality and safety) better with their own forces.

After the decisions have been made regarding which scopes of work to subcontract, subcontractors should be notified about the project. It is a good idea to have the subcontractors working to assist with the estimate in parallel with the general contractor developing its own estimates for self-performed work. A subcontractor call sheet similar to the one illustrated in Box 3.1 can be used to both call subcontractors and verify interest in estimating the project as well as to assist with developing individual subcontractors' scopes of work. When subcontractors are called (or today, most likely emailed), they will ask the estimator questions regarding specifications, quantities, and materials. The estimator should be somewhat informed at that time about specific subcontractor scope but should be cautious about providing too much detail to the subcontractor. General contractors do not want to place themselves in the position that subcontractors have based their prices solely on information the GC estimators have given them. Each subcontractor should develop a completely independent estimate. Incorporating the cost of subcontracted portions of the work into the project estimate is discussed more fully below.

Information obtained through review of all of the foregoing processes and documents will assist the estimator with preparation of the initial WBS. This WBS development will continue to evolve throughout the estimating, scheduling, and buyout processes that are all discussed later in this text. The first level of a work breakdown structure for the NanoEngineering Building is shown in Box 3.2. As the estimator dives into the project, there will be several more detailed subsequent levels of the WBS as well.

## Quantity take-off

The estimator takes material quantities directly off the drawings. The goal is to perform detailed measurements and counts of each item of work that has previously been selected to be self-performed. After each item has been taken off and recorded on quantity take-off (QTO) sheets, the drawings can be momentarily set aside. The take-off process is one of the most time-consuming building blocks in the estimating triangle shown in Figure 3.1 and is a critical step towards preparing the final total estimate value. Development of QTOs for subcontracted work by the GC is not as detailed as that of their own direct work. Order-of-magnitude (OM) estimates

## Box 3.1   Subcontractor call sheet

### SUBCONTRACTOR CALL SHEET

*NanoEngineering Building*
*2/10/2015*

**Sub/Supply Item:**   *Floor Covering*

**Specification (s):**   *09-30-00, 09-65-43, and 09-68-13*

| Potential Bidders: | Contacts: | Phone: | Will Bid? Yes or No |
|---|---|---|---|
| Omar's Flooring | Omar Smith | 214-1992 | Yes |
| Coyote Covering | Clancy Torrence | 214-4444 | No |
| Stein's | Randy Redfish | 214-9820 | Yes |
| United Tile | Terry Jones | 214-9829 | Tile Only |
| Soft Touch | Sherri Bradford | 214-0872 | Carpet Only |

**General Scope of Work:**   *All floor covering including carpet and tile and base*

**Specific Inclusions:**   *Carpet, pad, vinyl, rubber base, ceramic tile, stair treads, transition strips, adhesives*

**Specific Exclusions:**   *Floor prep by NWCC*
*We will need to unload their trucks with our forklift*

**Other Notes:**   *Also wainscot or plam counters? Will they bid union? Advantage to NWCC if we can get one sub for all work*

are initially developed for subcontracted work to be replaced later with cost estimates and quotes provided by the subcontractors.

Is it necessary to count every nail? No. Is it necessary to count every window? Yes. Eighty percent of the costs are included with 20 percent of the work items (80–20 rule). The experienced estimator will focus on the bulk of the cost and risk exposure, and the minor items will get covered.

The quantity take-off process starts with the work items that will be constructed first: the foundations. This will accomplish several tasks. First, it will force the estimator to think like

# Box 3.2    Work breakdown structure

## WORK BREAKDOWN STRUCTURE
*NanoEngineering Building*

| CSI Division | Description |
|---|---|
| 01 | Jobsite General Conditions |
| 02 | Demolition |
| 03 | Concrete: |
|  | Purchase Reinforcement Steel |
|  | Foundations |
|  | CIP Walls and Slabs |
|  | Post-Tension Elevated Slabs |
| 04 | Masonry:        CMU and Stone Veneer |
| 05 | Structural and Miscellaneous Steel |
| 06 | Wood and Plastic: |
|  | Rough Carpentry |
|  | Finish Carpentry |
| 07 | Thermal and Moisture Protection: |
|  | Waterproofing |
|  | Insulation |
|  | Roof and Accessories |
|  | Gutters and Downspouts |
| 08 | Doors, Windows, and Glass: |
|  | Door Frames and Leafs |
|  | Windows and Storefront |
|  | Door Hardware |
| 09 | Finishes: |
|  | Drywall |
|  | Painting |
|  | Acoustical Ceilings |
|  | Floor Covering: Carpet |
|  | Floor Covering: Ceramic, Vinyl, Base |
| 10 | Specialties |
| 11, 12, 13 | Equipment, Furnishings, Special Conditions |
| 14 | Conveying Systems, Elevators |
|  | Mechanical: |
| 21 | Fire Protection |
| 22 | Plumbing |
| 23 | HVAC and Controls |
|  | Electrical Systems: |
| 26 | Line Voltage: Power and Lights |
| 27, 28 | Low Voltage: Communications, Safety |
| 31 | Earthwork: Mass Excavation, Backfill, and Shoring |
| 32 | Exterior Improvements: |
|  | Paving |
|  | Walks and Miscellaneous Site work |
|  | Landscaping |
| 33 | Utilities: Water, Sanitary Sewer, Storm |

the builder: The floor system is built before the walls. This is also why it is important to have the superintendent involved in the estimating development. Organization of the estimate in this fashion will later assist with the schedule development (Chapter 4) and will aid with the development of the project cost control systems (Chapter 12). All the work items are taken off prior to pricing. Material quantities are recorded and later extended out and summarized on QTO sheets similar to the one shown in Worksheet 3.2.

## Worksheet 3.2   Quantity takeoff

### Quantity Sheet

#### Northwest Construction Company

Project:  *Middle School Lump Sum Bid*                Date: *2/15/2015*
Owner/Loc:  *Washington School District*             Estimator:  *TJ*
                                                      Est #:  *1*

*Div # 03: Continuous Footings*

| Ref | Description | Qty (ea) | L (ft) | W (ft) | H (ft) | LF Edge Frm | CF Concr | SF F. Grade | LF Rebar |
|-----|-------------|----------|--------|--------|--------|-------------|----------|-------------|----------|
| 1/S5 | 1' – 8" x 10" | | 52 | | | | | | |
| | | | 37 | | | | | | |
| | | | 26 | | | | | | |
| | | | 34 | | | | | | |
| | | | 4 | | | | | | |
| | | | 155 | | | | | | |
| | | | 4 | | | | | | |
| | | | 34 | | | | | | |
| | | | 4 | | | | | | |
| | | | 350 | 1.67 | 0.83 | 700 | 485 | 585 | 1,050 |
| | Subtotal: | | | | | | | | |
| | | | | | | | | | |
| | 2' 0" x 10" | | 12 | | | | | | |
| | | | 6 | | | | | | |
| | | | 12 | | | | | | |
| | | | 6 | | | | | | |
| | Subtotal: | | 36 | 2 | 0.83 | 72 | 60 | 72 | 108 |
| | Subtotal: | | | | | 772 | 545 | 657 | 1,158 |
| | | | | | | LF | CF | SF | LF |
| | | | | | | | | | |
| | Waste and Lap: | | | | | | x 1.05 | | x 1.1 |
| | | | | | | | = 572 CF | | = 1,274 LF |
| | | | | | | | /27 CF/CY | | x 1.043 |
| | Conversions: | | | | | | = 21 CY | | = 1,329# |
| | | | | | | | | | /2,000#/T |
| | | | | | | | | | = 0.7 Tons |
| | | | | | | | | | |

Good estimating habits include the following:

- The drawings should be marked up, whether done by hand with colored pencils and hi-liters or shown on the computer, indicating what items of work have been taken off. This will help to minimize errors.
- Sketches and assumptions should be developed and noted, most preferably on the QTO sheets.
- Quantities should be measured, extended, and summarized as they will be purchased. For example, concrete is initially measured in cubic feet, but it is purchased in CY. It does not do any good to summarize concrete in cubic feet on the pricing sheets if it will be purchased in CY. Similarly, many roofing systems are measured in SF but are purchased by the square (100 SF or CSF or 1 SQ).
- Waste factors for all quantities should be included on the quantity take-off sheets. Table 3.3 provides some commonly used waste factors borrowed from *Construction Cost Estimating, Process and Practices*.
- The estimate should appear neat and professional. Upper management will want to review the estimator's work prior to bid for completeness and accuracy. The neater the document, the more confidence it will exude.
- Although most estimating activities are being performed electronically today, paper print-outs are an inexpensive and useful method to quality-control check the estimate. Estimating errors can not only lose bids, they can bankrupt a contractor.
- A clear paper trail is important. The estimate will be used later for schedule development and cost control systems.

After all of the material has been taken off the drawings, the estimator recaps or groups like sizes and material products together in preparation for transfer onto the pricing pages. All areas of concrete are grouped and added together. In this way, only total yardage is brought forward

**Table 3.3** Material waste factors

| Common waste factors for cost estimating | |
|---|---|
| Concrete: | 5–7% for most work; 3% for large placements |
| Wire mesh: | 10% for lapping of mesh |
| Reinforcement steel: | 10% for lap and bends or 40% times bar diameter |
| Building paper: | 10% for lapping of paper and waste |
| Vapor barrier: | 10% for lapping of material and waste |
| Gravel: | 10 to 20% for additional volume due to compaction |
| Earth fill: | 20 to 30% for volume increase due to compaction |
| Framing lumber: | 10% for waste and backing and blocking |
| Plywood: | 10% for waste |
| Siding and roofing: | 20–30% for lap |

Source: Construction Cost Estimating, Process and Practices

to the pricing pages and in only one location. The purchase of other major materials, which may be recapped, include reinforcement steel and structural steel. Similarly, some contractors may add together all of the millwork or all of the rough carpentry, regardless of size. This may be done because the labor to install such materials will be similar or the same per ton or lineal foot or board foot. Other materials that may be grouped together include items such as doors and windows.

The amount of waste to apply will vary with the installer, project, and estimator. Allowing between 5 percent and 10 percent is common. Purchase of enough, but not too much, material is important to maintain labor productivity. Items such as nails, glue, caulking, and rebar tie-wire can be estimated or allowed for, but determining exact quantities is difficult and outside of our 80–20 rule. The amount of time and cost the estimator expends exceeds the value of the materials. Allowances are usually sufficient at this stage.

## Pricing self-performed work

Pricing recapitulation sheets (recaps) are developed for each system or Construction Specification Institute (CSI) division that is utilized. An example of a pricing recap sheet is shown in Worksheet 3.3. All of the recapped material quantities are brought forward and entered onto the pricing sheets. The estimator should not start pricing until all of the materials have been taken off the quantity sheets and entered on the pricing sheets, similar to the process of taking the materials off the drawings. Each quantity is circled or highlighted, indicating it has been brought forward. After every quantity has been taken off a quantity sheet, the sheet is marked, indicating that it has been completed. After all the quantity sheets have been brought forward, the estimator begins labor and material pricing.

Direct craft labor productivity is the most difficult item for a contractor to estimate and is therefore the riskiest. Often a general contractor will review the amount of labor in an estimate and use this figure as some basis for determining overall project risk and therefore the appropriate fee to apply. Labor productivity is estimated as man-hours/unit of work, such as 10 man-hours per door. This is referred to as unit man-hours or UMH. This system allows for fluctuation of labor rates, union versus open shop, geographic variations, and time. If it takes 0.8 man-hours per yard to pour a strip footing in Atlanta, it probably takes the same in Alaska. Appropriate wage rates can be applied for the specific project. The best source of labor productivity is from the estimator's desk drawer or in-house database. Each estimator or contractor should develop his or her own database from previous estimates and previously developed labor factors. Other sources of UMH would include published databases or reference guides, such as Means' *Building Construction Cost Data.*

It is important to round man-hours off. Fractional extended man-hours should not be retained on the pricing recap sheet. Partial man-hours are difficult to schedule and monitor against for cost control, let alone difficult to explain to the superintendent why he or she has 1.7 man-hours to place the gravel for the perforated drain system and not 2 hours. Estimating is not that accurate a science. The man-hours are totaled at the bottom of each pricing page. This figure will be valuable information later for scheduling and cost control.

The wage rates that get applied are determined by the company, location, union agreements—if applicable— and the type of work to be performed. Sources of wage rates may include Davis-Bacon rates that are established by the Department of Labor, prevailing wage rates, union rates, or in-house rates. An example of prevailing wage schedule for the NanoEngineering Building as

# Worksheet 3.3   Pricing recapitulation sheet

### NORTHWEST CONSTRUCTION COMPANY
*1242 First Avenue, Cascade, Washington 98202*
*(206) 239-1422*

| Middle School Bid | Owner: Washington School District | Estimator: Ted Jones | EST #1 |
| Lump Sum Estimate | Architect: ABC Partners | Date: 07/15/15 | SHEET: 1 |

| Description | Quantity | Unit | Labor | | | | Material | |
|---|---|---|---|---|---|---|---|---|
| | | | UMH | MH | Rate | Cost | UP | Cost |
| **SYSTEM:** | | **CONCRETE FOUNDATIONS PRICING RECAP** | | | | | | |
| Layout: 3 man crew | 1 | DAY | 24.00 | 24 | 38.00 | 912 | 100.00 | 100 |
| Excavate and export foundations | 185 | TCY | 0.10 | 19 | 32.00 | 592 | 15.00 | 2,775 |
| Import and place backfill | 96 | TCY | 0.10 | 10 | 32.00 | 307 | 22.00 | 2,112 |
| Perforated drain system: Pipe | 250 | LF | 0.05 | 13 | 32.00 | 400 | 2.50 | 625 |
| Gravel | 17 | TCY | 0.10 | 2 | 32.00 | 54 | 25.00 | 425 |
| Fabric | 700 | SF | 0.02 | 14 | 32.00 | 448 | 0.50 | 350 |
| Fine grade under foundations | 985 | SF | 0.05 | 49 | 32.00 | 1,576 | 0.25 | 246 |
| Form continuous footings | 772 | LF | 0.08 | 62 | 38.00 | 2,347 | 0.80 | 618 |
| Form spot footings | 280 | SF | 0.10 | 28 | 38.00 | 1,064 | 1.00 | 280 |
| Form pilasters | 180 | SF | 0.20 | 36 | 38.00 | 1,368 | 1.20 | 216 |
| Form elevator pit | 130 | SF | 0.20 | 26 | 38.00 | 988 | 1.20 | 156 |
| Rebar: See separate recap | 0 | | 0.00 | 0 | 40.00 | 0 | 0.00 | 0 |
| Pour continuous footings | 21 | CY | 0.80 | 17 | 32.00 | 538 | 10.00 | 210 |
| Pour spot footings | 11 | CY | 1.00 | 11 | 32.00 | 352 | 10.00 | 110 |
| Pour pilasters | 6 | CY | 2.00 | 12 | 32.00 | 384 | 10.00 | 60 |
| Pour elevator pit | 2 | CY | 2.00 | 4 | 32.00 | 128 | 11.00 | 22 |
| Buy concrete | 120 | CY | 0.00 | 0 | 0.00 | 0 | 115.00 | 13,800 |
| Pump concrete | 120 | CY | 0.00 | 0 | 0.00 | 0 | 15.25 | 1,830 |
| Concrete equipment | 120 | CY | 0.00 | 0 | 0.00 | 0 | 50.00 | 6,000 |
| Concrete accessories | 120 | CY | 0.00 | 0 | 0.00 | 0 | 15.00 | 1,800 |
| Tubesteel column embeds | 14 | EA | 1.00 | 14 | 38.00 | 532 | 25.00 | 350 |

| SUBTOTAL: | | | 339 Hours | | | $11,990 | | $32,085 |
|---|---|---|---|---|---|---|---|---|
| LABOR TAX: | | @50% of labor | | | | | | $5,995 |
| LABOR: | | | | | | | | $11,990 |
| TOTAL SYSTEM: | | | | | | | | $50,070 |
| SYSTEM COST $/CY: | | $50,070/120 CY = $417/CY | | | | Checks | | |
| AVERAGE WAGE RATE: | | $11,990/339 HR = $35.36 per hour | | | | Checks | | |

(System cost of $417/CY is high but many non-foundation items are included here and quantities are small, therefore acceptable)

well as the Seattle area unions is shown in Table 3.4. A measured material quantity of 6,500 SF of slab on grade requiring finishing is multiplied times the historical UMH of .01 MH/SF to yield 65 MH. The cement finisher's wage rate of $35 per hour is applied, yielding a total of $2,275 to finish the slab.

On non-prevailing wage projects, only bare (paid to the employee) wage rates are included on the pricing sheets. Labor burdens or labor taxes are included on the estimate summary page.

**Table 3.4**   Labor wage rates

| Craft | Union wage rate | Prevailing wage rate* |
| --- | --- | --- |
| Carpenter | $38/hour | $54/hour |
| Cement finisher | $35/hour | $54/hour |
| Drywall taper | $38/hour | $54/hour |
| Electrician | $46/hour | $65/hour |
| Floor covering/tile mason | $28/hour | $32/hour |
| Glazier | $40/hour | $56/hour |
| Ironworker | $40/hour | $64/hour |
| Laborer | $32/hour | $44/hour |
| Operating engineer | $40/hour | $56/hour |
| Painter | $29/hour | $39/hour |
| Plumber | $46/hour | $46/hour |
| Sheet metal worker | $47/hour | $73/hour |
| Teamster | $32/hour | $51/hour |

* Prevailing wage rates include labor taxes as dictated by jurisdiction. Additional labor benefits, such as training and retirement, may also be added depending upon location and union preferences.

This line item will add between 30 percent and 60 percent, depending upon craft, on top of the labor estimate for expenses to the general contractor for items such as workers' compensation, union benefits, unemployment, FICA (social security), and medical insurance. These percentages fluctuate with time and location. The estimator should check with his or her company's current accounting data before applying a percentage.

The best sources of material unit prices are suppliers. Suppliers are requested to provide prices for each item of material that will be purchased by the general contractor. Price quotations should be received in writing on the supplier's letterhead. If telephone quotations are received, forms such as the one illustrated in Box 3.3 should be used for recording material pricing, although pricing on the vendor's letterhead is always preferred. Many material quotes are received electronically. Other sources of material prices include historical costs or previous estimates. Historical prices are not as reliable as current quotes, but it is better than leaving an item blank. The third choice for material prices are databases or reference guides. These sources are useful for unusual items for which local quotes cannot be obtained or as allowances or subcontractor plugs or OM estimates in early budget development.

The material prices are extended by multiplying the quantity of 120 CY of concrete times the concrete supplier's quote of $115 per CY to yield a material price of $13,800. Extended prices should be rounded to the nearest dollar. Fractional man-hours and cents are applicable in the unit price or UMH columns of the pricing recap page only, not the extended columns. After all the unit prices have been applied to the quantities, totals are brought down and in preparation to be brought forward to the estimate summary page.

A good reality check of the estimate for a given system, say foundations, can be made at this time on the pricing recap page. In the lower right corner of the sheet, add together the labor cost,

# Box 3.3    Telephone quote

### NORTHWEST CONSTRUCTION COMPANY
1242 First Avenue, Cascade, Washington 98202
(206) 239-1422

| Middle School Bid | Owner: Washington School District | Estimator: Ted Jones | EST #1 |
|---|---|---|---|
| Lump Sum Estimate | Architect: ABC Partners | Date: 07/15/15 | SHEET: 1 |

| Description | Quantity | Unit | Labor UMH | Labor MH | Labor Rate | Labor Cost | Material UP | Material Cost |
|---|---|---|---|---|---|---|---|---|
| **SYSTEM:** | **CONCRETE FOUNDATIONS PRICING RECAP** | | | | | | | |
| Layout: 3 man crew | 1 | DAY | 24.00 | 24 | 38.00 | 912 | 100.00 | 100 |
| Excavate and export foundations | 185 | TCY | 0.10 | 19 | 32.00 | 592 | 15.00 | 2,775 |
| Import and place backfill | 96 | TCY | 0.10 | 10 | 32.00 | 307 | 22.00 | 2,112 |
| Perforated drain system: Pipe | 250 | LF | 0.05 | 13 | 32.00 | 400 | 2.50 | 625 |
| Gravel | 17 | TCY | 0.10 | 2 | 32.00 | 54 | 25.00 | 425 |
| Fabric | 700 | SF | 0.02 | 14 | 32.00 | 448 | 0.50 | 350 |
| Fine grade under foundations | 985 | SF | 0.05 | 49 | 32.00 | 1,576 | 0.25 | 246 |
| Form continuous footings | 772 | LF | 0.08 | 62 | 38.00 | 2,347 | 0.80 | 618 |
| Form spot footings | 280 | SF | 0.10 | 28 | 38.00 | 1,064 | 1.00 | 280 |
| Form pilasters | 180 | SF | 0.20 | 36 | 38.00 | 1,368 | 1.20 | 216 |
| Form elevator pit | 130 | SF | 0.20 | 26 | 38.00 | 988 | 1.20 | 156 |
| Rebar: See separate recap | 0 | | 0.00 | 0 | 40.00 | 0 | 0.00 | 0 |
| Pour continuous footings | 21 | CY | 0.80 | 17 | 32.00 | 538 | 10.00 | 210 |
| Pour spot footings | 11 | CY | 1.00 | 11 | 32.00 | 352 | 10.00 | 110 |
| Pour pilasters | 6 | CY | 2.00 | 12 | 32.00 | 384 | 10.00 | 60 |
| Pour elevator pit | 2 | CY | 2.00 | 4 | 32.00 | 128 | 11.00 | 22 |
| Buy concrete | 120 | CY | 0.00 | 0 | 0.00 | 0 | 115.00 | 13,800 |
| Pump concrete | 120 | CY | 0.00 | 0 | 0.00 | 0 | 15.25 | 1,830 |
| Concrete equipment | 120 | CY | 0.00 | 0 | 0.00 | 0 | 50.00 | 6,000 |
| Concrete accessories | 120 | CY | 0.00 | 0 | 0.00 | 0 | 15.00 | 1,800 |
| Tubesteel column embeds | 14 | EA | 1.00 | 14 | 38.00 | 532 | 25.00 | 350 |

| SUBTOTAL: | | | | 339 Hours | | $11,990 | | $32,085 |
|---|---|---|---|---|---|---|---|---|
| LABOR TAX: | | @50% of labor | | | | | | $5,995 |
| LABOR: | | | | | | | | $11,990 |
| TOTAL SYSTEM: | | | | | | | | $50,070 |
| SYSTEM COST $/CY: | | $50,070/120 CY = $417/CY | | | | Checks | | |
| AVERAGE WAGE RATE: | | $11,990/339 HR = $35.36 per hour | | | | Checks | | |

(System cost of $417/CY is high but many non-foundation items are included here and quantities are small, therefore acceptable)

material cost, and calculated portion of labor taxes to come up with a quick total cost for that system. Then divide that figure by the total predominant material quantity, such as 120 CY of concrete, to come up with a total unit price for that system or assembly. There are many available references that are used for conceptual estimating and budgeting against which this number can be checked. There is also the more relevant personal list of figures the estimator has on hand, such as $350 per CY for concrete foundations. If the calculation is in the range, the estimator is comfortable. If it is way off, say one half or double, then maybe an error occurred, and a double check should be made.

## Pricing subcontracted work

The best source of subcontractor pricing is subcontractors. They are the ones who will ultimately be required to sign a contract and guarantee performance of an established amount of work for a fixed amount of compensation. General contractors should estimate subcontractor work to develop pre–bid day budgets and to check the reasonableness of subcontractor bids. If one low bid of $940,000 for carpet comes in and one higher bid of $1.2 million is available, the general contractor can feel comfortable throwing the high bid out because maybe its own estimate was $950,000. The high bidder obviously either does not have the right scope or is too busy at this time. The reverse may be true on another system, if the general contractor's estimate was nearer the high bid. Once a reliable subcontractor price is received, the GC's estimate figures should be replaced with the subcontractor's price. The estimator should not assume that the firms who specialize in areas of floor covering are all in error and that his or her figure is the most correct one. A spreadsheet should be developed which aids in the analysis of subcontractor and supplier pricing. An example for the floor covering for the NanoEngineering Building is shown in Worksheet 3.4.

## Jobsite general conditions

Even if the general contractor is not required to turn in a construction schedule, it is still necessary to develop a rough schedule prior to preparing a general conditions estimate. Utilizing past experience, coupled with the estimated man-hours and with the superintendent's input, it is a fairly simple process to prepare a 20–line item schedule. It is not necessary to determine whether it will take exactly 400 work days to build the building. However, it is important to know that this structure will take approximately 20 months to construct (not 10 or 30) given the site conditions, project complexity, long lead material deliveries, anticipated weather, and available manpower.

There are many different uses of the term general conditions. Jobsite general conditions costs are project specific and are measurable. It was discussed earlier here that estimating direct labor costs is one of the most difficult and riskiest tasks of the construction estimating team. Estimating jobsite general conditions is also difficult and risky but is easier to quantify than estimating home office indirect costs that may be applicable to a project. Site specific general conditions could also be referred to as jobsite administration costs. These jobsite expenses will be based upon this specific project's duration, needs, and risks. A summary schedule for NanoEngineering is presented in Chapter 4.

An excerpt of a listing of typical jobsite general condition's items is shown in Worksheet 3.5. The complete general conditions estimate for this project is quite involved, including costs for a tower crane, and is available on the companion website. Commissioning is an important element of close-out on many projects that have complicated mechanical and electrical systems, such as a laboratory building. Commissioning and other close-out activities are discussed later in Chapter 17. Commissioning was a requirement of the NanoEngineering project, and NWCC has included associated costs in this general conditions estimate. There is not an exact rule for what the total general conditions costs should be. They could range anywhere from less than 5 percent for a larger project to greater than 10 percent for a smaller project. The distinction of which items might be jobsite reimbursable and which are part of the fee will be spelled out in the contract and will affect this range. Many of the individual general conditions line items are project- and site-specific, and others are time-dependent. As introduced, the NanoEngineering Building was a CM-

# Worksheet 3.4   Subcontractor bid spreadsheet

**Subcontractor and Supplier Bid Analysis**
*NanoEngineering Building*
*2/15/2015*

Bid Package:   *Floor Covering*

| Contractors: | NWCC Plug | Omar | Coyote | Stein | United | SoftTouch |
|---|---|---|---|---|---|---|
| Bid Per Spec? | | Alt. | No Bid | Yes | Yes | Yes |
| Union Labor? | | No | | Yes | Yes | Yes |
| Delivery Time: | | 2 Mo | | 60 Day | In-Stock | ?? |
| Exclusions: | Stair Tread | Floor Prep | | | Carpet | Vinyl |
| | | Stairs | | | | |
| Include Tax? | No | No | | No | No | No |
| Unit Pricing: | | Carpet $30/SY | | | | Base $1.5/LF |
| | | Tile $10/SF | | | Tile $11/SF | |
| Quantities? | | ?? | | | ?? | ?? |
| Notes: | Floor Prep | | | | | |
| | w/Concr. | | | | | |
| Addenda: | 3 | 2 | | 1 | 3 | 3 |
| Base Bid: | | $1,775,000 | | $1,575,000 | $750,000 | |
| Carpet | $950,000 | in | | in | excl | $940,000 |
| Base | $15,000 | in | | excl | excl | $14,000 |
| Vinyl | $700,000 | in | | in | in | excl |
| Tile | $135,000 | | | excl | in | excluded |
| Subtotal: | $1,800,000 | $1,775,000 | | $1,575,000 | $750,000 | $954,000 |
| Adjustments: | $0 | $0 | | $150,000 | NA | $750,000 |
| Adjusted Bid: | $1,800,000 | $1,775,000 | | $1,725,000 | | **$1,704,000** |

At-risk project where the jobsite general conditions were bid lump sum along with the project fee. The general conditions on that project were approximately 9 percent, which is higher than the standard range, but that project included a tower crane, Leadership in Energy and Environmental Design, building information models, dedicated quality control and safety personnel, full-time traffic control, and other costs particular to working on an active university campus.

Before the jobsite general conditions estimate can be developed, the anticipated labor hours must be obtained from the direct work portion of the estimate. The superintendent and project manager rough out a schedule based upon these hours, the project's complexity, and subcontractor

# Worksheet 3.5  General conditions estimate

| NanoEngineering Building | Owner: UW | Estimator: Ted Jones #1 | Estimate |
|---|---|---|---|
| GMP Estimate | Architect: ZGF | Date: 03/18/15 | Sheet: 1 of 1 |

| Description | Quantity | Unit | Labor | | Material | |
|---|---|---|---|---|---|---|
| | | | Rate | Cost | UP | Cost |
| **SYSTEM:  JOBSITE GENERAL CONDITIONS** | | | | | | |
| Project Manager | 99 | wks | 2,100 | 207,900 | 0 | 0 |
| Superintendent | 95 | wks | 2,400 | 228,000 | 0 | 0 |
| Assistant Superintendent | 31 | wks | 2,000 | 62,000 | 0 | 0 |
| Project Engineers: 2 @ 99 wks | 197 | wks | 1,300 | 256,100 | 0 | 0 |
| Commissioning | 50 | wks | 1,300 | 65,000 | 0 | 0 |
| Office Administrator | 23 | mos | 5,000 | 115,000 | 0 | 0 |
| Field Survey | 16 | wks | 4,000 | 64,000 | 4,000 | 64,000 |
| QC Inspector | 22 | mos | 8,000 | 176,000 | 0 | 0 |
| Off-Shift Security | 10 | mos | 0 | 0 | 7,500 | 75,000 |
| Police for Traffic Control | 15 | mos | 0 | 0 | 15,000 | 225,000 |
| In-house Flagger | 99 | wks | 1,000 | 99,000 | 0 | 0 |
| Safety Inspector | 22 | mos | 8,000 | 176,000 | 2,500 | 55,000 |
| Forklift Rental & OE | 20 | mos | 12,000 | 240,000 | 7,000 | 140,000 |
| Small Tools | 3,023,350 | DL$ | 0 | 0 | 0.04 | 120,934 |
| | | | | | | |
| Tower Crane:  Erect/Dismantle | 2 | LS | 0 | 0 | 50,000 | 100,000 |
| Tower Crane:  Rent | 13 | mos | 0 | 0 | 20,000 | 260,000 |
| Tower Crane:  Operator | 13 | mos | 1,000 | 130,000 | 0 | 0 |
| | | | | | | |
| Rubbish Removal | 22 | mos | 0 | 0 | 5,000 | 110,000 |
| Periodic Cleanup | 22 | mos | 8,000 | 176,000 | 0 | 0 |
| Final Cleanup | 1 | quote | 0 | 0 | 51,684 | 51,684 |
| Glass Cleaning | 1 | LS | 0 | 0 | 35,000 | 35,000 |
| Preconstruction Services | 1 | LS | 0 | 0 | 250,000 | 250,000 |
| SUBTOTAL LABOR AND MATERIAL: | | | | 2,194,081 | | 2,422,168 |
| LABOR: | | | | | | 2,194,081 |
| LABOR BURDEN: | 38% | | Average | | | 833,751 |
| TOTAL SYSTEM: | | | | | | $5,450,000 |
| Cost per month: | 22 | | | | | $247,727 |
| Percentage of direct cost: | | $56,761,237 | | | | 9.6% |

Percentage is high, but due to tower crane, security, traffic, safety, QC, etc. therefore acceptable.

input on durations and deliveries. This schedule is used to develop overall durations for estimating site administration and equipment rental type durations for the jobsite general conditions estimate.

The general conditions costs are not estimated until after the schedule is developed. A reasonable estimate of construction time is one of the most crucial elements to estimating general conditions. If the project lasts 22 months, then certain support individuals, equipment, and material will need to be on site for that duration.

## Company overhead indirect costs

Company home-office overhead indirect costs include items such as accounting, marketing, and officer salaries. These costs usually are a relatively fixed figure based upon staffing for 1 fiscal year. An average-size general contractor may need to generate 1.5 percent to 3 percent (that is, 1.5 percent to 3 percent of their annual volume) to cover these costs. In construction accounting, we refer to this as the *breakeven margin*. This markup is applied to the total estimated direct project cost and sometimes is called the *overhead burden*. In addition to covering the indirect costs, the contractor desires to earn a profit on each project. The desired profit plus the home office overhead indirect cost is called the *fee*. Fees can range anywhere from 3 percent to 8 percent for commercial work. Smaller commercial projects require higher fees due to smaller annual volumes of the firms involved. If the general contractor self performs much of the work, they also have a proportional increased risk and fee. On cost-plus contracts, conceptual cost estimates may be used to establish the guaranteed maximum price, but company overhead costs still need to be estimated to develop the fee proposal.

## Estimate summary wrap-up

In addition to choosing the fee, there are several other final steps in putting the estimate together. A one- or two-page estimate summary form will be utilized to gather all the direct work estimates, general conditions, subcontractor pricing, and markups. Subcontractor pricing that hasn't yet been received should be priced with in-house estimates. These in-house estimates often are referred to as "plugs." The pre–bid day summary is just a refinement of the first rough order of magnitude (ROM) estimate that was developed the day the documents came in. As the estimator completes the estimate, he or she should use the most current and relevant information. All the pricing recap pages are brought forward by posting the values on the summary sheet. If a certain project has specific needs that the estimate summary page does not account for, a project specific summary must be prepared.

Usually at the bottom of the estimate summary, below the subtotal of the direct costs, are often a series of percentage markups including the fee. Some of the other markups or add-ons that are applicable on many projects, be they bid or negotiated, include the following:

- Material tax: varies between states. Sometimes these values are already included in the pricing. Sometimes they apply only to materials that are not incorporated into the project, for example concrete forms.
- Contingency: If the project is competitively bid with relatively complete documents, the amount of contingency applied is usually zero. Most general contractors account for the contingency, if any, with their choice of fee on competitive projects. Stated contingencies will show up on negotiated projects that have incomplete documents.

- Insurance: Liability insurance is volume-related. The insurance rate will range significantly between contractors depending upon their size and safety records. The range generally ranges from less than 1 percent for large commercial contractors to 2 percent for smaller contractors.
- Business tax or excise tax: This is also job cost–related and is dependent upon the city, county, and state.
- Bond: Performance and payment bond rates are also annual volume- and company performance-related, but the project bond costs are project specific. They generally do not appear on negotiated projects. On larger lump sum commercial projects, they may be required, especially for public work, and the cost will range depending upon the size of the project and the past performance of the contractor. The bond cost is customarily an alternate add-on and not included in the bid. A sample bond rate schedule was shown in Chapter 2.

In the case of a bid job, the estimate summary should be filled out as much as possible the day prior to bid day. The estimator should have someone else check the math as it is here that many gross errors occur. An estimate summary form, similar to other estimating forms, should be developed and used consistently throughout the firm. It may require slight modification for any one project, but its consistency is important to provide the project estimating team a quick and comfortable overview prior to and during bid day. A hard copy should be printed out and saved. The computer is an excellent and standard tool to use at this time and throughout the estimating process.

The final bid or proposal value is determined on bid day. This is the top step in Figure 3.1 estimating triangle. Subcontractor quotations will be received, and the figures revised. The final bid generally is approved by an executive of the construction firm. The total figure must be submitted on the form specified in the instructions to bidders or proposers to the owner before the specified time.

As stated earlier, the client for our case study project utilized the CM-at-risk delivery method. The contractor's fee and pre-construction services and general conditions were competitively bid. After award, and with additional design information, the general contractor prepared a guaranteed maximum price estimate. Some of the details were known and estimated with firm prices, such as earthwork and concrete, and others were covered by plugs and allowances. This allows the contractor to begin construction work on a fast-track basis while the balance of the design is being completed. Bids were received by NWCC for many of the more expensive and therefore riskier subcontract areas prior to submission of the GMP to the client. The subcontracts were not awarded though until after the GC had received its contract from the client. An abbreviated GMP summary estimate is shown in Worksheet 3.6 and an expanded version is shown on the companion website. In a lump sum project, the summary estimate might be only one page, whereas a completely open book negotiated project would require several pages.

## Summary

Similar to constructing a building, estimating is a series of steps. The first is project overview to determine whether the project is going to be pursued. The quantity take-off step is a compilation of counting items and measuring volumes. Pricing is divided between material pricing, labor pricing, and subcontract pricing. Labor is computed using productivity rates and labor wage rates. Material and subcontract prices are developed most accurately using supplier and subcontractor quotations. Jobsite general conditions cost is a job cost and is schedule-dependent.

## Worksheet 3.6    Estimate summary

### SUMMARY ESTIMATE FOR NANOENGINEERING BUILDING
3/25/2015

| Line Item | Description | | Estimated Cost |
|:---:|:---|:---:|---:|
| 1 | Jobsite General Conditions | | $5,450,000 |
| 2 | Demolition | | $850,000 |
| 3 | Concrete | | $8,055,000 |
| 4 | Masonry | | $1,525,100 |
| 5 | Structural & Miscellaneous Metals | | $803,030 |
| 6 | Wood & Plastic | | $937,500 |
| 7 | Thermal & Moisture Protection | | $3,080,900 |
| 8 | Doors, Windows, Glass | | $2,673,000 |
| 9 | Finishes: | | |
| 9.1 | Drywall | | $1,778,800 |
| 9.2 | Painting | | $590,000 |
| 9.3 | Acoustical Ceilings | | $722,000 |
| 9.4 | Floor Coverings: Carpet, Ceramic, Vinyl | | $1,604,550 |
| 10 | Specialties | | $388,290 |
| 11-13 | Equipment, Furnishings, Special Construction | | $2,740,350 |
| 14 | Elevator | | $842,000 |
| 21-23 | Mechanical Systems: | | |
| 21 | Fire Protection | | $772,000 |
| 22 | Plumbing | | $2,750,000 |
| 23.1 | HVAC | | $10,500,923 |
| 23.2 | Controls | | $1,300,000 |
| 26 | Line Voltage Electrical | | $5,000,000 |
| 27 | Low Voltage Electrical | | $1,148,000 |
| 31 | Earthwork: Excavation, Backfill, Shoring | | $3,139,664 |
| 32 | Exterior Improvements | | |
| 32.1 | Paving | | $255,050 |
| 32.2 | Walks & Miscellaneous Sitework | | $500,000 |
| 32.3 | Landscaping | | $255,000 |
| 33 | Site Utilities | | $295,000 |
| | Subtotal Direct and Indirect Costs: | | $57,761,237 |
| | | | |
| | Estimating & Escalation Contingency @ 1%: | $577,612 | $58,338,849 |
| | Liability Insurance @ 0.75%: | $437,541 | $58,776,391 |
| | State Excise Tax @ 0.8%: | $470,211 | $59,246,602 |
| | CM Fee @ 3%: | $1,777,398 | $61,024,000 |
| | **Total Contract, Excluding State Sales Tax:** | | **$61,024,000** |

Home office indirect costs are often combined with desired profit to produce the fee. The fee calculation varies dependent upon several conditions including company volume, market conditions, labor risk, and resource allocations. The three major lessons to be learned in estimating is first to be organized. If proper organization and procedures are utilized, good estimates will result. The second is to estimate and estimate a lot. Practice and good organization will eventually develop thorough and reasonable estimates. The third is to maintain and use historical databases of cost and productivity data from previous projects.

## Review questions

1.  What is the difference between a conceptual cost estimate and a detailed cost estimate?
2.  What is the greatest risk the contractor faces when developing a cost estimate?
3.  How does an estimator develop the work breakdown structure for a project?
4.  When is a construction schedule developed in the estimating process and why?
5.  What is the difference between jobsite general conditions and home office indirect costs?
6.  What are two sources for labor productivity rates?
7.  List five of the general contractor's personnel who should participate in developing an estimate for the project and what their specific roles would be.
8.  Why is it important that the first conceptual estimate developed by a contractor be somewhat accurate?
9.  What is wrong with using unit prices from published references for material and subcontract items of work?
10. What is the difference between a construction schedule and an estimate schedule?
11. What is the 80–20 rule and how does it apply to estimating concrete formwork?
12. If an estimator is certain of receiving at least one or two subcontract bids on bid day, why should he or she bother with estimating that area of work themselves?
13. What is an acceptable jobsite general conditions percentage? Should straight percentage calculations be used in developing an estimate?
14. Where do most estimate errors occur?

## Exercises

1.  Using Figure 3.2 and the organization chart in Figure 1.1 from Chapter 1, develop an estimating assignment list for each team member, including due dates. Include a "check person" for each major deliverable.
2.  Complete and expand the project item list shown in Worksheet 3.1 for the NanoEngineering Building. Items that require decision regarding self-performed or subcontractor work should be noted.
3.  Complete and expand the WBS shown in Box 3.2 and on the website for the NanoEngineering Building to (1) the second level, and then (2) the third level.
4.  In addition to the data shown in Worksheet 3.6 summary estimate and on the website for NanoEngineering, what other detail would a construction-savvy client look for from a contractor on a completely open-book project? If you were the client, list your top ten items.

# 4

# Planning and scheduling

## Introduction

Like estimating, planning and scheduling have been covered extensively in other books dedicated solely to that topic. Our purpose in this chapter is to briefly discuss schedule development and then focus on using the schedule as a project management tool. As discussed in Chapter 1, time management is just as important to project success as is cost management. The key to effective time management is to carefully plan the work to be performed, develop a realistic construction schedule, and then manage the performance of the work. Schedules are working documents that need updating as conditions change on the project. For example, contract change orders generally modify some aspect of the scope of work requiring an adjustment to some part of the schedule. In addition to aiding management of the project, updated schedules also can be used to justify additional contract time on change orders and claims. Some texts use the terms *scheduling* and *project management* identically. This is a too all-encompassing use of the two terms. Project management, as can be seen within the context of this book, involves many other aspects of construction management than just scheduling.

## Project planning

Project planning is the process of selecting the construction methods and the sequence of work to be used on a project. Planning must be completed before a schedule can be developed. It starts with the assembly of all the information necessary to produce a schedule. Other steps in planning the project follow.

- Developing the work breakdown structure (WBS), which is a listing of all the activities that must be performed to complete the project; a first-level example WBS for the NanoEngineering Building was shown in Chapter 3.

- Acquiring input from the key members of the project team including:
  1. company specialists,
  2. field supervisors,
  3. subcontractors, and
  4. suppliers.
- Making decisions with field supervisors regarding:
  1. site layout,
  2. sequence in which work will be performed,
  3. direction of work flow: bottom up, left to right, east to west,
  4. means and methods of construction,
  5. type of concrete forming system to be used,
  6. material handling, equipment, and hoisting, and
  7. safety requirements.
- Making decisions regarding performing work with the general contractor's own work crews or with subcontractors.
- Identifying all restraining factors such as:
  1. skilled labor: crew makeup and sizes,
  2. material and equipment delivery date estimations,
  3. weather,
  4. permits, and
  5. financing.
- Preparing the estimate. The schedule is not prepared until the estimate has been completed, at least through the work that is to be performed by the contractor's direct work force. Craft hours from the estimate are needed to determine activity durations for the schedule. However, a preliminary schedule is needed to estimate the cost of jobsite general conditions, as was discussed in Chapter 3.

The project plan may be an outline, notes, minutes of meetings with responsible parties, or a roughed-out schedule. This information is then provided to the scheduler, who in some instances may be the same individual as the planner. On some projects, the project manager and/or the superintendent may do the planning and develop the schedule. The preconstruction phase includes estimating and scheduling and other forward-thinking planning processes as will be discussed in more detail in Chapter 5.

## Types of schedules

There are two primary types of schedules: bar charts and network diagrams. Bar charts relate activities to a calendar but generally show little to no relationship among the activities. Network diagrams show the relationship among the activities and may or may not be time-scaled on a calendar. Two techniques are used in developing network schedules. The first is known as the *arrow diagramming method* in which arrows depict the individual activities. The other is known as the *precedence diagramming method* in which the activities are represented by nodes. The arrow diagramming method used to be prepared manually, but today both the arrow and the precedence diagramming methods are computer-generated schedules. Network diagram schedules are sometimes referred to as *critical path* schedules because the critical path is the longest path through the schedule and determines the overall project duration. Any delay in any activity on the critical path results in a delay in the completion of the project. All are good systems and may be appropriate in different applications.

Schedules can be prepared in different formats depending upon the anticipated use.

- Formal schedules may be developed and provided to the owner as required by contract or prepared and submitted with a proposal for a negotiated contract.
- Summary schedules, like the one illustrated in Figure 4.1, often are used for presentations or management reporting.
- Detailed schedules are posted on the walls of meeting rooms or in the jobsite trailer. They are marked up with comments and progress. A sample detailed schedule for the NanoEngineering Building is included on the text companion website.
- Short-interval "look ahead" schedules are developed by each superintendent or foreman and each subcontractor each week. They may be hand-drafted in bar chart form, through Excel, or with scheduling software such as Microsoft Project. The form or system does not matter. What is important is that they are produced by the people who are doing the work and that they are communicated and distributed to all involved. In this way, they are construction management tools. These schedules can be two-, three-, or four-week increments, depending upon the job and level of activity. Worksheet 4.1 shows an example of a 3-week startup schedule from our case study project. As indicated, the schedule format is not as important as its author and content. The superintendent for the NanoEngineering Building prepared his initial schedule utilizing Excel.
- Mini-schedules, area schedules, and system schedules allow additional detail for certain portions of the work that could not be adequately represented in the project schedule and have longer duration than the short-interval schedule.
- Pull planning schedules are lean construction tools being adapted from production industries and are discussed in Chapter 5.
- Other specialized schedules include submittal schedule, buyout schedule, material delivery schedule, equipment startup schedule, and close-out schedule, some of which are discussed in other chapters in this text.

## Schedule development

Once the project plan has been completed, it is time to develop a construction schedule. Individual tasks or activities to be accomplished were identified during the development of the WBS. The next step is to determine the sequence in which the activities are to be completed. This involves answering the following questions for each activity:

- What deliverables, permits, submittals, and contracts need to be in place before starting?
- What activities must be completed before this activity can start?
- What activities can be started once this activity has been completed?
- What activities can be performed concurrently with this activity?

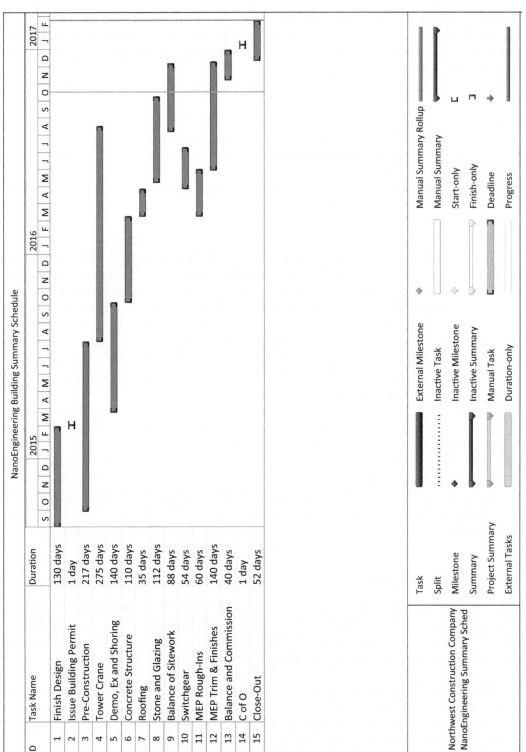

**Figure 4.1** Summary schedule

# Worksheet 4.1   Short interval schedule

## SHORT INTERVAL SCHEDULE

NORTHWEST CONSTRUCTION COMPANY
PROJECT:          NanoEngineering Building
SUPERINTENDENT:   Jim Bates

DATE: 8/25/15
SHEET 1 OF 1

| NO. | ACTIVITY DESCRIPTION: | S | S | 23 | 24 | 25 | 26 | 27 | S | S | 30 | 31 | 1 | 2 | 3 | S | S | 6 | 7 | 8 | 9 | 10 | S | S | COMMENTS |
|---|---|---|---|---|---|---|---|---|---|---|---|---|---|---|---|---|---|---|---|---|---|---|---|---|---|
| | | | | | | March 2015 | | | | | | | | April 2015 | | | | | | | | | | | |
| 1 | Notice to Proceed | | | | | X | | | | | | | | | | | | | | | | | | | Original in Office |
| 2 | Deliver Office Trailers | | | | | | X | | | | | | | | | | | | | | | | | | |
| 3 | Install Silt Fence | | | | | | | X | | | X | X | | | | | | | | | | | | | |
| 4 | Survey Corners | | | | | | | X | | | X | | | | | | | | | | | | | | |
| 5 | Locate Utilities | | | | | | | | | | X | | | | | | | | | | | | | | |
| 6 | Install Temp Power | | | | | | | | | | | | | X | X | | | | | | | | | | |
| 7 | Deliver Sanikan | | | | | | | X | | | | | | | | | | | | | | | | | |
| 8 | Clear and Grub Site | | | | | | | | | | | | | | | | | | | | X | X | | | Finish Monday |
| 9 | Precon Meeting with City | | | | | | | | | | X | | | | | | | | | | | | | | |
| 10 | Mobilize Demolition Sub | | | | | | | | | | | X | | | | | | | | | | | | | |
| 11 | Demo Existing Structure | | | | | | | | | | | | X | X | X | | | X | X | X | X | | | | |
| 12 | Geo Approve Import Sample | | | | | | | | | | | | X | X | X | | | X | X | X | X | X | | | Expedite! |
| 13 | Utility Relocates | | | | | | | | | | | | | | | | | | | | | | | | Post Clear & Grub |

Based on the answers to these questions, the schedule structure can be developed using the interdependencies among the activities. Even with all the computer sophistication available today, many contractors still develop their logic by hand on butcher paper or by rearranging colored stickie-notes on the whiteboard. A photograph was taken from an actual example of this from the NanoEngineering jobsite meeting room and is also posted on the companion website. The duration for each activity is determined using the crew productivity factors used in developing the cost estimate. Now the start and finish dates for each activity can be determined, and the overall scheduled project completion date is compared with the contractual completion date. If the schedule shows completion later than the contract requires, some of the activities on the critical path must be accelerated or stacked to produce a schedule that meets contractual requirements.

Computers are valuable tools to assist in developing the schedule. The identification of the activities, the duration for each, and the sequence in which they are to be completed must be input into the computer for it to develop the schedule. The computer will plot the schedule and calculate the start and finish time for each activity. Computer-generated schedules allow the scheduler and project manager and superintendent to determine quickly the effects of changing schedule logic, delays in delivery of critical materials, or adjusting resource requirements. The computer is a valuable scheduling tool, but it cannot take the place of adequate planning.

The project manager and superintendent should both be actively involved in developing the construction schedule, for they will have the responsibility for completing the project in the desired time. This provides buy-in by the project leadership. If they agree with the schedule, they will do everything they can to make it happen.

## Schedule updates

In order for a schedule to be an effective construction management tool, it must be used and not just printed out and posted on the wall. It should be annotated weekly to monitor progress in relation to scheduled work. The superintendent customarily reports progress at the weekly owner/architect/contractor (OAC) meetings. He or she often ties a plumb-bob (or in the case of the NanoEngineering Building, a 1-inch-diameter nut (see website photograph) to a string and moves this status line each week during the meeting. Hi-liters can be used to visually indicate activities that are ahead or behind. Scheduling software programs also have means to depict schedule progress. The overall schedule status and any important issues should be documented in the weekly meeting notes. Meeting notes are discussed in Chapter 10.

Should the project manager ever update or revise a schedule? If so, when and how? This depends upon both the contract requirements as well as project progress. If the project is proceeding more or less on schedule, the schedule need not be updated. If the project is significantly behind schedule or there have been many change orders, the schedule should be updated. Current computer scheduling software allows revisions to the schedule to be saved. This provides the capability to compare the current schedule with the original schedule allowing the project manager to document schedule impacts when negotiating change orders.

The original schedule should be maintained intact if at all possible. If a revised schedule is developed, it must be submitted to all subcontractors to be evaluated for impact. Upon acceptance, the new schedule must be incorporated into every contracting and supplying firm's respective contracts. Similar to incorporating new drawings into contracts, this is a change order opportunity for subcontractors.

Schedules, like estimates, provide a map of where the project team wants to go and how they are going to get there. The objective is not necessarily to complete each activity exactly within its scheduled duration but to complete the overall project within the scheduled time. Many activities will be completed early, and others late. The project manager and superintendent just want to keep from finishing the entire project late. The objective is for the variances to average out.

Sometimes it is necessary to revise the schedule, either because of a contract requirement or because of a substantial change in scope or progress. It is recommended that some system be used to retain the original commitments of all of the parties. If it appears that construction progress is significantly behind, it may be necessary to develop a new schedule. This may have been caused by a variety of reasons including:

- late material deliveries,
- owner-induced delays,
- architect induced delays,
- weather,
- additional scope,
- labor shortages or inefficiencies,
- equipment availability,
- inadequate estimate or schedule, or
- incorrect schedule logic.

Just as the project manager and superintendent should not work off an inaccurate cost estimate, they should not use an incorrect construction schedule. The construction team must first understand what the problems are and then endeavor to correct them. This may be done through a variety of methods including:

- paying extra for expedited material deliveries, such as air-freight,
- increasing manpower,
- working double shifts,
- working overtime,
- changing craftsmen or field supervisors, and
- removing and replacing a subcontractor or supplier.

Most of these are not without increased costs and other impacts and risks, including production inefficiencies. The project team must carefully assess the impacts of each method selected. Schedule recovery usually can occur only by accelerating activities on the critical path.

## Cash flow analysis

A cash flow curve is a projection of the total value of the work to be completed each month during the construction of the project. It is created by cost-loading the schedule and plotting the total monthly costs. Often, it is one of the first things the owner will ask of the project manager and may be specifically required by the construction contract. One reason this is required is to provide information to the bank for anticipated monthly payments. Some project managers resist on the basis that the curve will be wrong and that they may be penalized for it. The most important thing a project manager does is to get paid from the owner for the work that has been completed on the project. This will be discussed in more detail in Chapter 11. If a cash flow curve is a requirement to facilitate payment, it should be developed.

The cash flow curve is easy to prepare. The first step is to develop a cost-loaded schedule. The estimated costs are applied across the construction schedule activities. If an activity will take 5 months and its value is $100,000, then $20,000 is spread over each of the 5 months. Costs such as administration, taxes, and fees should be distributed proportionately over the entire project. The cost data are summed at the bottom of the schedule for each month to develop anticipated monthly expenses. The likelihood of the general contractor being billed by each material supplier and subcontractor according to any anticipated schedule is somewhat remote. It would be the exception to the rule, but the project manager may be limited on monthly payment requests by these cash flow projections. The contract requirements should be reviewed to determine whether there are any restrictions. Worksheet 4.2 shows a contractor's cost-loaded schedule for NanoEngineering Building. Due to space limitations this is a condensed quarterly version of the 2-year monthly analysis included with the companion website.

Now a cash flow curve can be plotted. This can be displayed in either bell shape or s shape. The bell-shaped curve represents a plot of the estimated value of work to be completed each month. The s-shaped curve represents a plot of the cumulative value of work completed each month. Some project managers will adjust the monthly figures to reflect somewhat a standard bell curve. Within reason, this is acceptable for presentation purposes but is not a requirement. Many cash flow curves actually depict more of a double-hump camel than a bell. This is caused when there are significant project costs early on the project, such as pre-payment for long-lead equipment, and late in the project, such as expensive finishes. Our case study is a research building, and significant expenses are realized when laboratory casework commences. The actual cash flow can later be tracked against the schedule. Figure 4.2 shows the contractor's work-in-place cash flow curve from the Worksheet 4.2 cost-loaded schedule.

One interesting twist to the cash flow analysis is that there are actually several different means of measurement:

- Committed costs: The purchase orders have been issued, and the subcontracts have been awarded. Therefore, the project manager and the owner have committed to spend the money, but it may not yet be paid.
- On-site materials: The reinforcement steel was received, but the project manager has not received an invoice for it, and therefore, has not made payment.
- Costs in place: The light fixtures that were delivered last month are installed this month. Costs for materials are not usually counted until the materials are actually installed on the project, especially if installed by a subcontractor.
- Costs billed to the general contractor: This reflects invoices received from suppliers and subcontractors. It usually lags behind the costs committed and in place.
- Monthly pay request to the owner: This follows receipt of invoices from suppliers and subcontractors. Payment can be up to 30 days after some of the labor was paid and materials have been received.

# Worksheet 4.2   Contractor's cost loaded schedule

**COST LOADED SCHEDULE**
8/15/2015

$ × 1,000:

| Item | Description | Cost | 2nd Qtr 2015 | 3rd Qtr 2015 | 4th Qtr 2015 | 1st Qtr 2016 | 2nd Qtr 2016 | 3rd Qtr 2016 | 4th Qtr 2016 | 1st Qtr 2017 | Totals |
|---|---|---|---|---|---|---|---|---|---|---|---|
| 2 | Demolition | $850,000 | 850 | | | | | | | | $850,000 |
| 3 | Concrete | $8,055,000 | | | 4,200 | 3,855 | | | | | $8,055,00 |
| 4.1 | CMU | $450,100 | | | | | 450 | | | | $450,100 |
| 4.2 | Stone Veneer | $1,075,000 | | | | | 400 | 675 | | | $1,075,000 |
| 5 | Structural & Misc Metals | $803,000 | | | | 600 | 203 | | | | $803,000 |
| 6.1 | Rough Carpentry | $457,500 | | | | 458 | | | | | $457,500 |
| 6.2 | Finish Carpentry | $480,000 | | | | | | 240 | 240 | | $480,000 |
| 7.1 | Insulation | $835,000 | | | | | 170 | 500 | 165 | | $835,000 |
| 7.2 | Roof & Accessories | $1,095,000 | | | | | 1,095 | | | | $1,095,000 |
| 7.3 | Waterproofing | $575,000 | | | 150 | 425 | | | | | $575,000 |
| 7.4 | Sheetmetal Siding & Flashing | $575,980 | | | | | 576 | | | | $575,980 |
| 8.1 | Doors | $752,000 | | | | | 127 | 375 | 250 | | $752,000 |
| 8.2 | Windows & Storefront | $1,522,000 | | | | | 800 | 722 | | | $1,522,000 |
| 8.3 | Door Hardware | $399,000 | | | | | 69 | 200 | 130 | | $399,000 |
| 9.1 | Drywall | $1,778,800 | | | | | 670 | 670 | 439 | | $1,778,800 |
| 9.2 | Painting | $590,000 | | | | | | 300 | 290 | | $590,000 |
| 9.3 | Acoustical Ceilings | $722,000 | | | | | | 250 | 472 | | $722,000 |
| 9.4 | Floor Covering: Carpet | $954,000 | | | | | | 400 | 554 | | $954,000 |
| 9.5 | Floor Covering: Ceramic, vinyl | $750,550 | | | | | | 300 | 451 | | $750,550 |
| 10 | Specialties | $388,290 | | | | | | 200 | 188 | | $388,290 |
| 11 | Equipment, Furnishings | $2,740,350 | | | | | | 1,500 | 1,240 | | $2,740,350 |
| 14 | Conveying Systems (Elevator) | $842,000 | 200 | | | | | 500 | 142 | | $842,000 |
| 21 | Fire Protection | $772,000 | | | | 100 | 270 | 270 | 132 | | $772,000 |
| 22 | Plumbing | $2,750,000 | | 500 | 500 | 250 | 750 | 750 | | | $2,750,000 |
| 23 | HVAC & Controls | $11,800,923 | | | | 1,200 | 3,500 | 3,500 | 3,601 | | $11,800,923 |
| 26 | Electrical Systems | $5,000,000 | | 500 | 500 | 500 | 1,500 | 1,300 | 700 | | $5,000,000 |
| 27 | Low-Voltage Electrical | $1,148,000 | | | | | | 500 | 648 | | $1,148,000 |
| 31.1 | Excavation & Backfill | $1,999,000 | 500 | 1,400 | 99 | | | | | | $1,999,000 |
| 31.2 | Shoring | $845,664 | 100 | 600 | 146 | | | | | | $845,664 |
| 32.1 | Paving | $255,050 | | | | | | 200 | 55 | | $255,050 |
| 32.2 | Walks & Misc. Sitework | $500,000 | | | | | | 500 | | | $500,000 |
| 32.3 | Landscaping | $255,000 | | | | | | | 255 | | $255,000 |
| 33 | Site Utilities | $295,000 | 50 | | | | | 200 | 45 | | $295,000 |
| | Subtotal Direct Costs: | $52,311,237 | $1,700 | $3,000 | $5,595 | $7,388 | $10,580 | $14,052 | $9,997 | $0 | $52,311,237 |
| | Jobsite General Conditions | $5,450,000 | 710 | 710 | 710 | 710 | 710 | 710 | 710 | 480 | $5,450,000 |
| | Fee $ % Markups | $3,262,763 | 136 | 210 | 356 | 458 | 638 | 834 | 605 | 26 | $3,262,763 |
| | **Qtr Total, Excluding Tax** | **$61,024,000** | 2,546 | 3,920 | 6,661 | 8,555 | 11,928 | 15,596 | 11,312 | 506 | **$61,024,000** |
| | **Cumulative Totals:** | $61,024,000 | 2,546 | 6,466 | 13,127 | 21,682 | 33,610 | 49,206 | 60,518 | 61,024 | $61,024,000 |

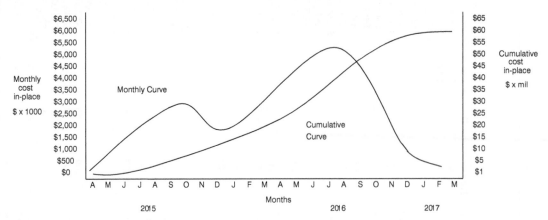

**Figure 4.2** Contractor's work-in-place curve

- Payment received: This reflects when payment was received by the contractor from the owner and the owner from the bank. It will lag behind the formal monthly payment request by 10 to 30 days.
- Payment distributed: This curve reflects payments distributed to subcontractors and suppliers. It will generally lag behind payment received from the owner by an additional 10 to 30 days.

So, which measure of cash flow should the curve represent? In most instances, the project manager will develop the curve based upon his or her schedule of construction that reflects the anticipated costs in place. This was the method used developing the curve shown in Figure 4.2. The formal invoice will be 1 to 4 weeks behind this curve, and the receipt of cash and subsequent disbursement to subcontractors up to a total of 2 months behind the time when the work was accomplished. In this way, the actual cash flow will fall behind the projection. An example cash flow analysis for the client is shown in Figure 4.3, that, when compared to Figure 4.2, depicts the 30-day delay in receipt of the client's monthly payment. Utilizing the earlier work-in-place curve is beneficial to all parties. The owner's and/or the bank's actual outflow of cash will be slower than had been scheduled. The only time a problem can occur is when the outflow of cash by the contractor is faster than had been scheduled and the money is not yet available from the owner or bank.

## Project management issues

The schedule should be referenced in the contract as an exhibit. It should be referred to in each subcontract and purchase order by title and date. The project manager may insert language into all subcontracts placing the subcontractors on notice that they are required to achieve the project schedule and that if the general contractor determines that they are falling behind, the subcontractors will work overtime and/or increase crew sizes at their expense to catch up.

Day 1 for the project schedule for the general contractor should hinge upon four things from the owner:

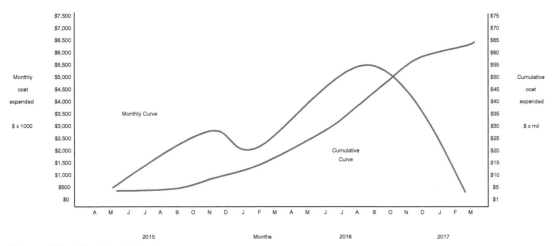

**Figure 4.3**   Client's cash flow curve

- verification of adequate construction financing,
- receipt of all building permits,
- receipt of a signed contract, and
- receipt of a notice to proceed (NTP).

These four requirements are often inserted into the contract. For our example, ConsensusDocs contract 500, it would be Article 6 unless modified by amendment. Most owners and general contractors will avoid using a specific date in either the commencement or completion articles of the construction contract. This was discussed more extensively in Chapter 2. Completion is usually defined as so many, say 600 calendar days or 400 work days, after issuance of the NTP and the other aforementioned requirements.

Performing construction on a "fast-track" basis was a buzzword of the 1980s. Most construction projects are fast-track now. Basically this means that activities are overlapping and occurring in parallel in lieu of in series. Very few schedules are completely linear in that the preceding activity is 100 percent complete before the subsequent activity starts.

Activity float represents the flexibility available to schedule activities not on the critical path without delaying the overall completion of the project. Who owns the float? It is owned by the first party who uses it. The owner may claim the float is theirs and use it to introduce new scope. The architect may use some of the float when responding to requests for information and submittals. Subcontractors and suppliers may use some of the float and either start their work late or deliver materials late. There have been numerous debates over float ownership, including many resolved by the courts.

Similar to cost control, the schedule should be developed with sufficient detail that it can be adequately measured, monitored, and corrected if necessary. Too much detail and the schedule will take on a life of its own and become too burdensome to monitor. The 80–20 rule should be applied to schedules in the same way it is applied to estimates, filing, and cost control work packages: 80 percent of the work or time is included in 20 percent of the activities. The project team should focus on that critical 20 percent.

Does the general contractor schedule the subcontractors, or do the subcontractors schedule themselves? The same principle of commitment applies to the subcontractor-developed schedules as it does to superintendent-developed estimates and schedules. If the subcontractor develops or, at a minimum, inputs to the schedule, and if it fits within the overall plan, they have made a commitment to achieve the schedule. Whereas, if the general contractor dictates to the subcontractors when and how long they will be on the job, the subcontractors may accomplish the task, but if they don't finish on time, they will always have the excuse that they did not get the opportunity to provide input.

Schedules should be maintained in a manner similar to the way record drawings and as-built estimates are maintained. They should be marked up and commented upon to reflect when activities started, when they finished, when materials were delivered, rain days, crew sizes, and whatever else the project manager and superintendent feel is appropriate. The as-built schedule, similar to the daily job diary, becomes a legal tool in the case of a dispute. The recording of actual durations can also be used to more accurately develop future schedules and estimates.

Unfortunately, many schedules today are used largely for claim preparation. On a recent project, a sizable subcontractor refused to provide any schedules of any value during the entire 2 years they were on the project. Remarkably, 2 months after completion, the subcontractor submitted a claim for extra costs incurred due to schedule delays. An exhibit to that claim was an incredibly detailed schedule, in color, computer-generated, with floats, deliveries, and manpower restraints all indicated, which had not been shared while the job was in progress. The reason this happened was because the document did not exist until after the fact; it was developed by a professional claims consultant. This schedule was not a construction management tool. If it had been available early in the project, maybe the subcontractor would have managed their work properly and communicated early with the general contractor, and a claim would not have been required.

## Summary

Schedules are important tools of all members: for the owner, design, and construction team. Proper planning of the project and the schedule, with input from the relevant personnel such as the superintendent and the subcontractors, is key to developing a useful construction management tool. Schedule development begins with proper planning that considers many variables such as deliveries, logic, manpower, and equipment availabilities. There are many different types of schedules, each of which has a use on a construction project. Updating or revising the schedule can be an expensive task and does not need to be a monthly occurrence, if the project remains on track. Using the schedule to monitor progress is the superintendent's responsibility and occurs during the weekly OAC coordination meeting. Development of a cash flow curve is a simple task and one that the project manager should do to assist the owner and the bank with analyzing financing requirements.

## Review questions

1. What is the difference between scheduling and planning?
2. Which members of the construction team input to the project plan?

3.  How many items should be included in the work breakdown structure?
4.  What is the difference between a bar chart schedule and a network diagram schedule?
5.  Who should develop a 3-week schedule and why they?
6.  What is the difference between updating the schedule and using it to monitor progress?
7.  Who indicates progress on the schedule? When should it be done?
8.  Why should contractors avoid revising and re-issuing the schedule?
9.  Are schedules contract documents?
10. Why do project managers resist developing cash flow curves?
11. Should a cash flow curve always be a perfect bell-shape curve?
12. When does day 1 of the contract schedule occur?
13. How many activities should be on the project schedule?
14. How does the 80–20 rule apply to developing schedules?
15. When is it appropriate to revise the schedule?
16. How is an as-built schedule developed, and when should it be developed?
17. Why does the client's cash flow curve take a slight turn up at the end of the project?

## Exercises

1.  Develop a short-interval schedule for the construction of the foundation work for the NanoEngineering Building. Identify the activities and logic and estimate reasonable durations for the activities.
2.  Utilizing the detailed schedule on the website, develop a schedule unique to the plumbing contractor. Include work of other trades as it affects the plumber's work.

CHAPTER

# 5

# Preconstruction planning

## Introduction

In this chapter, we will discuss the many activities that the project manager and superintendent undertake prior to starting the construction of a project. Some of these tasks may be required by the contract, while others may be undertaken to ensure a successful project. In the construction manager-at-risk project delivery method, the project owner may choose to execute a preconstruction services contract with the contractor. In the design-build project delivery method, preconstruction services are performed as part of the design-build contract. If the project design is completed before the general contractor is selected, such as in the design-bid-build project delivery method, there is no opportunity for the contractor to perform preconstruction services.

All construction contracts present risks to the general contractor, and the project manager and superintendent need to develop a risk management plan for each project. The allocation of risks between the project owner and the contractor is defined by the terms and conditions of the contract. Risk identification and management is a critical project manager responsibility. Value engineering (VE) may be performed as a preconstruction service or may be performed as a consequence of a VE incentive provision in the contract.

Successful projects require collaboration among many parties representing the project owner, the designer, the general contractor, the subcontractors, the suppliers, and the regulatory agencies. Some project owners include in their contracts provisions for establishing a partnering relationship among these parties. Building information models are being required by many project owners to provide tools to facilitate collaboration in the design and construction of their projects. Many owners are establishing sustainability goals for their projects that need to be considered in preconstruction planning.

Many contractors are adopting lean construction practices to minimize waste in project execution. One of the techniques being adopted for some projects is off-site construction to prefabricate project components that are later installed in the project.

## Preconstruction services contracts

The project owner may choose to select the construction contractor during the development of the project design and ask the contractor to perform preconstruction services. Such services may include:

- consultation,
- preliminary cost estimating,
- preliminary project scheduling,
- constructability analysis,
- value engineering, and
- site logistics planning.

Consultation involves attending design coordination and review meetings and providing advice regarding the use of materials, systems, and equipment and cost and schedule implications of design proposals. Preliminary cost estimates are developed using conceptual cost-estimating techniques and refined as the design is completed to ensure the estimated cost of the project is within the owner's budget. A preliminary schedule may be developed to assess the time impacts of design alternatives. Constructability analysis is reviewing the proposed design for its impact on the cost and ease of construction. VE (which is described in more detail later in this chapter) seeks to find the most economical building components from a life-cycle cost perspective. Site logistics planning involves analysis of construction site access and staging sequences.

Preconstruction services are more prevalent on privately funded projects than they are on public projects. This is because private owners are not restricted to the public bidding procedures that are required of many public owners. Contractors may negotiate a lump sum contract with an owner for preconstruction services or a time and materials contract. An example preconstruction services contract for the NanoEngineering Building is shown in Box 5.1. The owner must first define the set of preconstruction services desired and then negotiate a service contract with the construction contractor. Some owners use construction management firms to perform preconstruction services, if the owners are not able to select construction contractors before the designs are completed. Some public agencies use this approach. Owners hire contractors or construction managers to perform preconstruction services to provide construction expertise during design development to minimize cost, schedule, and constructability issues prior to bidding and during construction. A preconstruction services contract is a professional services contract similar to a design services contract and is not a construction contract.

## Risk analysis

As discussed in Chapter 2, construction is a risky business, and a construction contract is a mechanism for transferring some project risks from the owner to the contractor in return for some type of payment. To be successful, the project manager and the superintendent must identify the potential project risks and select strategies for managing them. The potential consequences of these project risks may pose threats to project success, but by using a proactive approach to identifying the risks and potential consequences of their occurrence enables selection of strategies for mitigating the effect on project success. This approach to developing a risk management plan is shown in Figure 5.1.

## Box 5.1    Preconstruction services agreement

### PRECONSTRUCTION SERVICES AGREEMENT

This agreement made this 16th day of December, 2014 between

the **Owner**:                          University of Washington
                                          Seattle, Washington 98195

and the **Contractor**:                  Northwest Construction Company
                                          1242 First Avenue
                                          Cascade, Washington 98202

for the following project:  NanoEngineering Building

The **Owner** and the **Contractor** agree as follows:

1.  That during the development of the design and prior to the start of construction, the **Contractor** will provide preconstruction services as follows:

*   Attend weekly coordination meetings.
*   Prepare budget estimates at the completion of conceptual documents, design development documents, during construction document development, and as otherwise required.
*   Develop cost analyses of design options.
*   Conduct value engineering studies as necessary to achieve budget goals.
*   Meet with consultants and/or subcontractors and suppliers, as necessary, to assist in the development of the design.
*   Conduct constructability review at completion of design development documents, 90" construction documents, and as otherwise required.

2.  That preconstruction services will be billed at the following rates:

|                          |                   |
| ------------------------ | ----------------- |
| Project manager          | $75 per hour      |
| Project superintendent   | $80 per hour      |
| Estimator                | $65 per hour      |
| Project engineer         | $50 per hour      |

3.  **Owner** will pay the **Contractor** the agreed-upon sum of $250,000 for preconstruction services. The **Contractor** will be paid monthly for its services based on actual time and expenses without markup, not-to-exceed the total compensation set above. Any amount that exceeds the total compensation agreed to will be at the **Contractor's** sole cost unless there are scope changes authorized by change order.

4.  All costs incurred by the **Contractor** under this contract will be billed on a monthly basis for payment, by the **Owner**, by the 10th of the following month.

5.  That the above agreement is hereby acknowledged and shall serve as the preconstruction services agreement between the **Owner** and the **Contractor**.

| Northwest Construction Company | University of Washington |
| ------------------------------ | ------------------------ |
| *Sam Peters*                   | *Jeffrey Jackson*        |
| Sam Peters                     | Jeffrey Jackson          |
| Vice President                 | Director, Capital Projects |

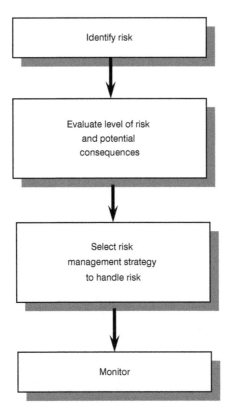

**Figure 5.1**   Risk management

As indicated in Chapter 2, the first level of risk analysis occurs when the contractor decides whether to pursue the project. While some business risks can be transferred by the purchase of insurance coverage, most project risks are not transferable. Many general contractors transfer the risks associated with specific scopes of work to subcontractors; however, the general contractors retain responsibility for the timeliness and quality of the subcontracted work. Selection and management of subcontractors will be discussed in Chapter 6. A tool for identification and management of potential project risks is shown in Worksheet 5.1. A project risk register can be used by the project manager and superintendent to identify potential project risks and craft strategies for managing them. Some of the risks typically encountered by many projects are crew productivity, material price escalation, quality control, weather, environmental requirements, equipment breakdown, subcontractor performance, public owner social goals, and safety. Safety will be addressed in Chapter 14.

## Value engineering

VE is a systematic evaluation of a project design to obtain the most value for the cost of construction. It includes analyzing selected building components to seek creative ways of performing the same function as the original components at a lower life-cycle cost without sacrificing reliability,

Worksheet 5.1    Project risk register

## Project Risk Register

Project: NanoEngineering Building                                          Date: January 7, 2015

| Risk Statement | | Risk Analysis | | Risk Response Strategy | |
| --- | --- | --- | --- | --- | --- |
| **Description of Risk** | **Construction Phase in which Risk may occur** | **Risk Probability** | **Risk Impact** | **Risk Response** | **Trigger** |
| Public safety during initial site utility construction | Initial site construction | High | Reduced productivity and potential public injuries | Install all site utilities including street and sidewalk trenching at night | Sitework subcontractor starting work |
| Erection of tower crane | Start of structure construction | High | Traffic impact of movement of crane components and erection with hydraulic crane | Install the tower crane on Saturday for minimum traffic interference with crane erection | Completion of site and foundation work |
| Traffic control during equipment and material deliveries | Throughout the project | High | Local pedestrian and vehicle traffic interfering with equipment and material deliveries | Hire retired Seattle police officer to control pedestrian and vehicle traffic enabling timely delivery of equipment and materials | Occupation of the project site |
| Cold weather concrete placement | Winter construction of concrete structure | High | Delayed placement of concrete components or cold temperatures during curing | Provide heaters and thermal blankets to protect concrete if cold temperatures occur | Cold temperatures during concrete placement |
| Late delivery of critical materials | Any phase requiring contractor purchased materials | Moderate | Delay in construction | Identify long-lead materials and place purchase orders early | Supplier notification of time required for delivery |

performance, or maintainability. VE studies may be performed by consultants during design development, as a contractor-performed preconstruction service, or by the contractor during construction. The most effective time to conduct such studies is during design development. Some lump sum construction contracts contain a VE incentive provision that allows the contractor to share in the savings that results from approved VE change proposals. VE change proposals submitted by the contractor are reviewed by the designer and owner for acceptability. If approved, up to 50 percent of the savings in construction cost may go to the contractor. The percentage split between the owner and the contractor will be stated in the VE engineering incentive provision contained in the contract. In a cost-plus contract, the savings may be shared if the final project cost is less than the guaranteed maximum price contract.

The value of a component or system can be defined as its function plus quality divided by its life-cycle cost. The life-cycle cost is:

initial or construction cost + operating cost + maintenance cost + replacement cost – any salvage value

VE studies are conducted to select the highest value design components or systems. These studies generally are conducted using the following five-step process:

- information gathering,
- speculation through creative thinking,
- evaluation through preliminary life-cycle costing,
- development of technical solutions, and
- presentation of alternative options.

The information-gathering phase involves studying the design to identify potential components or systems for detailed study. The essential functions of each component or system are studied to estimate the potential for value improvement. The VE study team needs to understand the rationale used by the designer in developing the design and the assumptions made in establishing design criteria and selecting materials and equipment.

The purpose of the speculation or creative phase is to identify alternative ways to accomplish the essential functions of the items selected for study. The intent is to develop a list of alternative materials or components that might be used. No intent is made to evaluate the identified alternatives but to generate ideas that will be evaluated in the next step of the study process.

The evaluation phase involves identifying the most promising alternatives from the set identified in the speculation phase. Preliminary cost data are generated, and functional comparisons are made between the alternatives and the design components being studied. The intent is to determine which alternatives will meet the owner's functional requirements and provide more value to the completed project.

The development phase involves developing design concepts for the alternatives identified during the evaluation phase. This involves developing detailed functional and economic data for each alternative. Estimated life-cycle cost data are developed for each alternative and compared with the estimated life-cycle cost of the components under study. The advantages and disadvantages of each alternative are identified. Alternatives are compared, and the ones representing the best value are selected for presentation to the designer and the owner.

The final step is the preparation of the VE proposals. Detailed technical and cost data are developed to support the recommendations. The advantages and disadvantages of each recommendation are described. The proposals are submitted to the designer and the owner for approval. If approved, the proposals are incorporated into the design. If not approved, the design is not changed. VE proposals approved after the construction contract is awarded must be incorporated into the contract by change orders. A simple approval does not provide the project manager the authority to deviate from the requirements of the contract plans and specifications.

## Partnering

Successfully executing a construction project involves the participation of many people representing numerous organizations. The ability of the project manager to produce a successful project depends on close working relationships among project participants. These contractual relationships generally are defined by contract language that many have viewed as creating adversarial situations. The result of this lack of teamwork often is seen in a significant number of contract disputes and resulting litigation. Owners, designers, contractors, and subcontractors are beginning to recognize that there is a better way to deliver projects. By working together in a cooperative attitude, many potentially strained relationships can be avoided, and participants can walk away from a completed project with a good feeling that they have been successful.

Partnering is a cooperative approach to project management that recognizes the importance of all members of the project team, establishes harmonious working relationships among them, and resolves issues in a timely manner minimizing impact on project execution. The team focuses on common goals and benefits to be achieved collectively during project execution and develops processes to keep the team working toward these goals. Partnering does not eliminate conflict but assumes all participants are committed to quality and will act in good faith toward issue resolution. Successful partnering requires an attitude shift among project participants. They must view the project in terms of making it a collective success rather than from their narrow parochial perspectives. Mutual trust and commitment are the trademarks of a successful partnering relationship.

Partnering has been used successfully to build cooperative relationships between the owner, the designer, and the contractor on projects. Project results, however, often were unsatisfactory because major subcontractors were not included in the partnering structure. Subcontractors often perform the bulk of the work on a construction project and have a great impact on project cost, duration, and quality. Major subcontractors must be included in the partnering structure if a cohesive project delivery team is to be forged.

The basic components of a partnering relationship are:

- a collectively developed mission statement,
- a collectively developed charter that contains specific goals and objectives,
- an effective communication system,
- an effective monitoring and evaluation system, and
- an effective issue resolution system.

Partnering will work only when there is total commitment from all participants to mutual trust and open, frequent communications. This commitment must originate with the leaders of all organizations represented on the project team. Unless there is commitment and support from the top, mutual trust will not permeate down into each organization represented on the project. This requires clear articulation regarding partnering intentions and a willingness to participate in initial partnering activities and succeeding periodic evaluations.

The decision to partner on a project usually rests with the owner. Such provisions would appear in the special conditions of the contract. Even though the owner has the lead, the contractor must be involved in the initial planning to ensure that all critical team members, including major subcontractors, are included. Sometimes critical suppliers and building inspectors also are included. Project success depends upon the inclusion of all key members of the project team.

The Special Conditions of the contract for construction of the NanoEngineering Building contained the following provision:

*The Owner and Contractor agree to use the partnering concept for this Project. Partnering empha-sizes a cooperative approach to problem-solving involving all key parties to the Project: Owner, Architect, Contractor and principal Subcontractors. Two workshops to define partnering relation-ships will be scheduled not-to-exceed one day each or as mutually agreed. The Owner, Architect and the Contractor will participate in one partnering session during preconstruction as soon as practicable. The Owner, Architect, Contractor, and principal Subcontractors, when known, shall participate in a second partnering session. Principal subcontractors should include: electrical, mechanical, sheet rock and others as the Contractor and Owner jointly agree are appropriate. The purpose of the workshops shall be:*

- *To establish mutual understanding of partnering concepts,*
- *To develop the mission statement and goals for the Project for all parties, and*
- *To develop a process so that critical issues can be quickly resolved.*

*The Owner will be responsible for providing the facilities for the workshops, as well as a facilitator and any workshop materials. The Contractor shall pay one-third of the cost for the facilitator and facilities not-to-exceed $2,000. The Contractor is expected to provide key Project personnel for the workshop at no additional cost to the Owner.*

The partnering process begins with gaining a commitment to the strategy from top management from each participating organization. Once the commitment has been made, a workshop is conducted by an external facilitator who is not involved in the project. An effective workshop requires careful planning. Both the owner's and the contractor's project managers should be involved. Specific issues to be addressed during workshop planning are as follows:

- What are the results desired from the partnership?
- How is top management support to be obtained?
- What is the budget for the workshop, and who pays the cost?
- Who should attend?
- Who is to be the facilitator?
- Does the facilitator understand the project and major concerns regarding its completion?
- Do attendees understand how partnering works and their responsibilities for making it successful?

- Where will the workshop be conducted?
- What will the workshop agenda be?
- Will the facilitator be involved in follow-up evaluation sessions?

Workshop attendees should include senior managers from all organizations represented as well as all personnel involved in managing activities on the project site. Senior personnel are needed to emphasize the commitment of each organization to the partnering relationship and ultimately to making the project a success for all participants. The objectives of the workshop are:

- for project participants to get to know one another,
- for all participants to understand the project success criteria of each participant,
- to identify potential problems and mitigation measures,
- to begin team building,
- to develop a partnering charter containing a mission statement and collective goals, and objectives for the project,
- to establish effective lines of communication,
- to develop a system for timely resolution of issues, and
- to establish an evaluation system.

A team-developed charter is fundamental to a successful partnering relationship. It defines the project team's mission and identifies collective goals that they established to define project success. The project goals should be specific, measurable, action-oriented, and realistic. Once the partnering charter has been crafted, it is signed by all workshop attendees, as illustrated in the partnering charter for the NanoEngineering Building project shown in Box 5.2.

The communication systems to be used on the project must be responsive to the needs of all project participants. A meeting schedule needs to be developed, and participants need to be identified. Specific media should be selected for keeping all project participants informed of project issues and their resolution. Meetings, telephone conference calls, and electronic mail generally are more responsive than formal correspondence. Time frames for responding to inquiries should be established.

An issue resolution system must be developed for timely escalation of open issues to minimize the impact on project execution. This system should delegate the initial attempt at issue resolution to the lowest levels in participating organizations. Issues that cannot be resolved within a specified time frame are then elevated to the next level for resolution. Each level is provided specific timelines until the open issue is elevated to the principals of the affected organizations. An example issue resolution system is shown in Box 5.3. The following rules are suggested for using the system:

- Resolve problems at the lowest level.
- Unresolved problems will be escalated upward by any party in a timely manner.
- No jumping levels of authority.
- Ignoring the problem or "no decision" is not acceptable.
- Do not make decisions you feel uncomfortable with; escalate unresolved issues upward.

Periodic evaluation of project performance is essential to monitor the success of the partnering relationship. Monthly performance reviews may be attended by project management personnel,

# Box 5.2    Sample partnering charter

## Partnering Charter

*Mission Statement*

We the partners of the NanoEngineering Building Project commit to combine our strengths and expertise as a team to construct a quality project safely in an atmosphere of mutual trust and respect achieving the following goals:

*Performance Goals*

- Excellent safety performance by completing the project with no fatalities, lost-time accidents, or public liability claims over $5,000.
- Completing a quality project that is built right the first time in conformance with the design intent.  Built like it was ours.
- Completing the project on time through timely resolution of issues and joint management of the schedule.
- Maximizing cooperation to limit cost growth to 2", minimize contractor and subcontractor costs, and minimize paperwork.
- Submittals reviewed and approved or approved with comments within 14 days.
- No unresolved claims at time of substantial completion.

*Communications Goals*

- Positive working relationship; be courteous, cordial, honest and listen.
- Open communications between all levels.
- Issue resolution at lowest level.
- Focus on issues and not personalities.
- Have productive meetings.
- Act responsibly.
- Develop a sense of pride, enthusiasm, integrity, and enjoyment on the project.  Have fun.

*Partner Signatures*

| | |
|---|---|
| *Robert Smith* | *Mary Peterson* |
| *Norm Riley* | *Jerry Brown* |
| *Sam Peters* | *Jack Alexander* |
| *Ted Jones* | *Alice Jackson* |
| *Linda White* | *Ernie Sanchez* |
| *Jim Bates* | *Barry Smith* |

## Box 5.3   Example issue resolution system

| Designer | Owner | Contractor | Time |
|---|---|---|---|
| Managing Partner | Executive Vice President | Chief Executive | 2 days |
| Project Sponsor | Director, Capital Projects | Vice President | 1 day |
| Project Manager | Project Manager | Project Manager | 4 hours |
| Design Coordinator | Project Engineer | Superintendent | 4 hours |
| Designer | Inspector | Foreman | 2 hours |

with principals attending quarterly reviews. Progress toward accomplishment of the charter goals and the effectiveness of the communications and issue resolution systems should be reviewed. Project executives should publicly state their commitment to making the relationship work and support their project personnel. Partnering successes should be recognized at these evaluation meetings to reinforce the value all place on maintaining the relationship. Project participants, not just managers attending formal evaluation meetings, should be asked to evaluate the relationship periodically.

## Team building

For the project team to succeed, project participants must agree to work collaboratively as a team to share responsibility for completing the project successfully. Since many of the participants may have never worked together previously, the first task is to get them to know one another. This is essential if mutual trust is to develop. It will not happen in one meeting, for developing trusting relationships takes time. Team members must learn to support one another and to collaborate freely and frequently. This generally is a greater problem on bid projects than it is on negotiated projects where team members may have worked together previously. Team building is the process of bringing together a diverse group of individuals and seeking to resolve differences, remove impediments, and proactively develop the group into a focused, motivated team that strives to accomplish a common mission.

Many construction projects are not managed in this manner. Unfortunately, information may be withheld for later use, and project participants may posture for their own advantage. Team members must establish guidelines that they will use in their team meetings. They must recognize that there may be conflicts among team participants but that they must not take any issue personally. All team members should be encouraged to state their views, and issues should be addressed objectively. Effective communication—listening, presenting, and discussing—is at the heart of teamwork. The team needs to discuss issues collaboratively, seek responsive solutions, and adopt a shared sense of accountability for achieving their collective goals and objectives. Effective teams are characterized by open communication, mutual trust, concern, support, and respect. This results from working together to solve mutual problems in a non-threatening, supportive environment.

# Building information models

Building information models were introduced in Chapter 1. They are useful tools for planning the execution of a construction project through information sharing with designers, subcontractors, and suppliers. They enable visualization of the project during the various phases of construction as well as when it is completed. The models can be used to determine constructability conflicts among the various disciplines involved in the design, such as structural, architectural, and mechanical. The various files are compared with one another to identify conflicts. The models also enable visualization of the work to reduce uncertainty, improve safety, resolve scheduling issues, and plan the use of prefabricated components.

The special conditions of the contract for the construction of the NanoEngineering Building contained the following provision:

> *The Project partners, including Contractor, will use Building Information Modeling (BIM) as a tool for collaboration, information sharing, estimating, planning and coordination. Contractor's direct cost for its BIM management program, including BIM integrator, shall be reimbursable as a Negotiated Support Service. Subcontractor cost for management and participation in BIM shall not be included in Negotiated Support Services and shall be included in the subcontract bid packages.*

As indicated in Chapter 1, BIM involves the creation of a coherent system of computer models that provide digital representations of the physical and functional characteristics of a project. Some of the models may be created by the designers, while others may be created by the general contractor and the subcontractors.

These models are used for project team collaboration as well as construction planning and management. The models can be viewed on mobile devices to support construction management in the field. Three-dimensional models of the construction site can be used to develop site utilization plans, including crane selection and placement, to minimize adverse impacts on construction operations. An example was shown in Figure 1.9 in Chapter 1.

Four-dimensional (4D) models, in which time is the fourth dimension, can be used to simulate construction operations for trade coordination, jobsite safety planning, and creation of pull planning construction schedules. Project superintendents can use the 4D models to simulate the sequence of concrete form construction as well as the placement of concrete. This will enable the superintendent to plan the most efficient sequence of work. The superintendent can also simulate the erection of the basic structure and crane placement to determine the optimal sequence of work. The 4D models can also be used to schedule material deliveries and identify temporary material storage areas on the site.

Another use of the model is to design building components that can be prefabricated off-site and installed on-site as assemblies. Such components include:

- structural assemblies,
- roof components,
- curtain walls and fenestration,
- stair assemblies,
- mechanical, electrical, and plumbing systems, and
- building components such as bathrooms.

BIM files contain parametric modeling information, and many fabricators use 3D models to build their components. Use of off-site construction techniques is discussed later in this chapter. Another use of the models is to support site layout. Some mechanical contractors are using building information model data to lay out their duct and pipe hanger support systems with lasers. An example model of mechanical systems interfacing with building structural elements is shown in Figure 5.2. Project managers need to plan for the use of BIM technology during preconstruction planning. Such technology can be very useful in monitoring the status of construction.

## Sustainable construction

Sustainable construction was introduced in Chapter 1. The objective is to plan and execute a construction project in such a manner as to minimize the adverse impact of the construction process on the environment. While much of the sustainable aspects of a completed project are a result of design decisions, other aspects are due to the manner in which the construction is performed.

Many project owners are interested in constructing sustainable buildings and seek to obtain certification from the US Green Building Council (USGBC) under its Leadership in Energy and Environmental Design (LEED) building assessment system. There are different LEED rating systems for various building types and projects. We will discuss the LEED v4 for Building Design and Construction (BD+C) system, which is for new construction and major renovations and was used on the NanoEngineering Building.

The LEED v4 for BD+C standard allows a maximum of 110 points distributed across eight categories as shown in Worksheet 5.2.

**Figure 5.2**   Building information model of mechanical work

# Worksheet 5.2  LEED v4 for bd+c project checklist

| Location and Transportation | | 16 Possible Points | Earned |
|---|---|---|---|
| Credit | Sensitive Land Protection | 1 | |
| Credit | High Priority Site | 2 | |
| Credit | Surrounding Density and Diverse Uses | 5 | |
| Credit | Access to Quality Transit | 5 | |
| Credit | Bicycle Facilities | 1 | |
| Credit | Reduced Parking Footprint | 1 | |
| Credit | Green Vehicles | 1 | |
| **Sustainable Sites** | | **10 Possible Points** | |
| Prerequisite | Construction Activity Pollution Prevention | Required | |
| Credit | Site Assessment | 1 | |
| Credit | Site Development – Protect or Restore Habitat | 2 | |
| Credit | Open Space | 1 | |
| Credit | Rainwater Management | 3 | |
| Credit | Heat Island Reduction | 2 | |
| Credit | Light Pollution Reduction | 1 | |
| **Water Efficiency** | | **11 Possible Points** | |
| Prerequisite | Outdoor Water Use Reduction | Required | |
| Prerequisite | Indoor Water Use Reduction | Required | |
| Prerequisite | Building-Level Water Metering | Required | |
| Credit | Outdoor Water Use Reduction | 2 | |
| Credit | Indoor Water Use Reduction | 6 | |
| Credit | Cooling Tower Water Use | 2 | |
| Credit | Water Metering | 1 | |
| **Energy and Atmosphere** | | **33 Possible Points** | |
| Prerequisite | Fundamental Commissioning and Verification | Required | |
| Prerequisite | Minimum Energy Performance | Required | |
| Prerequisite | Building-Level Energy Metering | Required | |
| Prerequisite | Fundamental Refrigerant Management | Required | |
| Credit | Enhanced Commissioning | 6 | |
| Credit | Optimize Energy Performance | 18 | |
| Credit | Advanced Energy Metering | 1 | |
| Credit | Demand Response | 2 | |
| Credit | Renewable Energy Production | 3 | |
| Credit | Enhanced Refrigerant Management | 1 | |
| Credit | Green Power and Energy Offsets | 2 | |

| Materials and Resources | | 13 Possible Points | Earned |
|---|---|---|---|
| Prerequisite | Storage and Collection of Recyclables | Required | |
| Prerequisite | Construction and Demolition Waste Management Planning | Required | |
| Credit | Building Life-Cycle Impact Reduction | 5 | |
| Credit | Building Product Disclosure and Optimization – Environmental Product Declarations | 2 | |
| Credit | Building Product Disclosure and Optimization – Sourcing of Raw Materials | 2 | |
| Credit | Building Product Disclosure and Optimization – Material Ingredients | 2 | |
| Credit | Construction and Demolition Waste Management | 2 | |
| **Indoor Environmental Quality** | | **16 Possible Points** | |
| Prerequisite | Minimum Indoor Air Quality Performance | Required | |
| Prerequisite | Environmental Tobacco Smoke Control | Required | |
| Credit | Enhanced Indoor Air Quality Strategies | 2 | |
| Credit | Low-Emitting Materials | 3 | |
| Credit | Construction Indoor Air Quality Management Plan | 1 | |
| Credit | Indoor Air Quality Assessment | 2 | |
| Credit | Thermal Comfort | 1 | |
| Credit | Interior Lighting | 2 | |
| Credit | Daylight | 3 | |
| Credit | Quality Views | 1 | |
| Credit | Acoustic Performance | 1 | |
| **Innovation** | | **6 Possible Points** | |
| Credit | Innovation | 5 | |
| Credit | LEED Accredited Professional | 1 | |
| **Regional Priority** | | **4 Possible Points** | |
| Credit | Regional Priority: Specify Credit | 1 | |
| Credit | Regional Priority: Specify Credit | 1 | |
| Credit | Regional Priority: Specify Credit | 1 | |
| Credit | Regional Priority: Specify Credit | 1 | |

| | | |
|---|---|---|
| **Total** | **110 Possible Points** | |

The level of certification gained is a function of the number of points earned by the project. The requirements for the four levels of certification are:

- certified (40–49 points),
- silver (50–59 points),
- gold (60–79 points), and
- platinum (80–110 points).

The special conditions of the contract for the construction of the NanoEngineering Building contained the following provision:

> *The LEED goal for this Project is a minimum of LEED Silver certification. The Contractor shall work collaboratively and proactively throughout all Phases of the Project, including construction to achieve this goal. The Contractor shall manage environmental issues and implement and document the Project's LEED requirements, including but not limited to: a) outline Subcontractor requirements for LEED in the subcontract bid forms; b) monitor the submittal process to ensure LEED compliance; c) train Subcontractors in LEED requirements; d) review design changes during construction for LEED impacts and notify Owner of impacts; e) ensure installed products are LEED compliant; and f) assemble and maintain records to document LEED goal compliance.*

Certification of a building starts with the project owner's decision to pursue LEED certification early in the design process. As part of the registration process, the owner establishes goals for the project in each of the eight categories. The LEED score sheet contains both prerequisite requirements as well as credit categories. Prerequisites are conditions that must be successfully addressed for a building to be eligible for consideration for a LEED rating. The sum of the credits earned is used to determine the level of certification.

The USGBC publishes a *Guide to LEED Certification* (www.usgbc.org/cert-guide) to provide guidance on the preparation and submission of the needed documentation and *LEED Project Checklist* (www.usgbc.org/resources/leed-v4-building-design-and-construction-checklist) to provide an electronic form for documenting the prerequisite and credit performance of the building.

Early in preconstruction planning, the general contractor must consider LEED requirements in the selection of materials, subcontractors, and construction strategies. Material reuse minimizes construction waste and earns materials and resources credits. Documentation will be needed to demonstrate achievement of certain credits, and submission of needed documentation must be included in supply contracts and subcontracts. Typical submittals required at this time are:

- site preservation and use plan,
- waste management and recycling plan, and
- indoor air quality plan.

Some of the standards in the *materials and resources* are under the control of the general contractor. There is a prerequisite for storage and collection of recyclables. There are credits for:

- construction waste management,

- materials reuse (building life-cycle impact reduction),
- recycled content in materials used (sourcing of raw materials), and
- use of certified wood (sourcing of raw materials).

Usually, a project engineer is assigned the responsibility for collecting the documentation and assembling the materials needed for submission to a third-party certification agent to validate achievement of the identified credits. The estimated cost of the project engineer's time to manage the LEED documentation needs to be included in the general contractor's general conditions cost estimate discussed in Chapter 3. As part of the material submittal process described in Chapter 7, the general contractor usually is required to provide the following:

- salvaged and refurbished materials,
- recycled content materials,
- regional materials, and
- certified wood products.

There is a LEED requirement for building commissioning as part of the contract closeout. This is discussed in Chapter 17. Building commissioning provides the project owner with a level of assurance that the building will function as designed.

## Environmental compliance

Depending upon the location of the construction project site, there may be multiple environmental restrictions placed on construction operations. There may be noise restrictions at night, or the site may be near a protected body of water or wetlands. In most instances, the general contractor will be required to control soil erosion and storm water runoff. An early submittal required by the project owner would be a project-specific storm water pollution prevention plan (SWPPP). This plan identifies potential sources of storm water pollution on the construction site and identifies measures to be implemented to eliminate polluted storm water from leaving the site. This often means taking steps to capture the storm water or retain it to enable infiltration into the soil. The quantity of soil erosion is influenced by the climate, topography, soils, and vegetative cover.

A typical table of contents for a SWPPP is:

- site assessment,
- erosion and sediment control measures,
- good housekeeping practices,
- inspections, and
- record keeping.

The SWPP should rely on erosion control as the primary means of preventing storm water pollution. Mats, geotextiles, and erosion control blankets may be used. Sediment controls provide a necessary secondary means of controlling storm water pollution runoff. Silt fences are often used as sediment control measures. The plan should address measures to be taken to control storm water flowing onto and through the project site, stabilize soils on site, protect storm drain inlets, and retain sediment on site.

The project may include demolition that may involve removal of materials that contain hazardous waste, such as lead-based paint or asbestos. Proper documentation and disposal requirements need to be understood by all parties involved in the removal of the hazardous waste. Spill prevention plans are often required to reduce the potential for contaminating the soil due to construction operations, such as fueling equipment. Many hazardous materials may be used in construction operations, and any excess hazardous materials must be disposed of properly.

## Lean construction

As introduced in Chapter 1, lean construction is a continuous process for analyzing the delivery of a construction project to minimize costs and maximize value. During preconstruction planning, the project manager and superintendent should examine planned processes for document flow, work flow, and material flow to minimize time delays and project costs. Web-based collaboration sites are frequently used for efficient management of project documentation to expedite review and coordination.

Work flow is managed by use of pull planning schedules that start with project milestones and work backward to define what work must be completed to achieve each milestone. Each week the project team reviews the work accomplished in the previous week and plans the work for the following week. Each foreman is asked to identify the work tasks that his or her crew will complete in the following week. Each individual foreman's commitment is connected to the commitments of the other foremen, thus creating a network of commitments among the project team. The result is better short-term planning and control by having work flow between crews without interruption. Pull planning does not replace the traditional construction schedules discussed in Chapter 4 but is used to improve the delivery of short-term assignments. The milestones selected to manage the pull planning scheduling process come from the traditional master construction schedule.

Pull planning schedules are often developed on white boards using multi-colored sticky Post-It notes as shown in Figure 5.3. The schedules are also used to manage the delivery of construction materials to ensure that all needed materials are available for the crews but that excess materials are not ordered to be stored on the construction site.

## Off-site construction

Another process for improving construction productivity is the use of off-site construction and prefabrication of building components and modules. Off-site construction reduces material waste, improves worker safety, provides better quality control, reduces trade interference, improves worker productivity, and reduces on-site construction time. Expanded use of BIM has facilitated the increased use of off-site construction. Many electrical and mechanical systems are being prefabricated in specialized shops and delivered to project sites as assemblies to be installed in projects under construction instead of being constructed piece by piece on site. This eliminates requirements for craft labor to use ladders to perform work on the project site, improving productivity and safety. In other cases, multi-craft modules are constructed off site that are delivered to the construction site for installation reducing cost and schedule and improving quality and safety. An example showing prefabricated mechanical racks is shown in Figure 5.4. The modules and prefabricated components

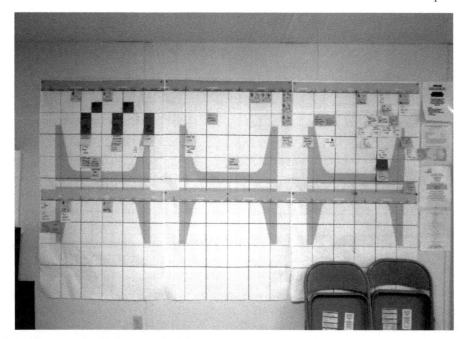

**Figure 5.3** Three-week pull planning schedule

**Figure 5.4** Installing prefabricated mechanical racks

need to be designed, and the designs need to be submitted to the project designers for approval. The size of the modules and components will be limited to the ability to transport them to a project site. Use of off-site construction may affect the project owner's cash flow for the project because components are fabricated well in advance of their installation on site so contractors need to ensure that owners are aware of the payment schedule.

## Summary

Project owners may ask the general contractor to perform preconstruction services while the design is being completed. The types of services typically requested are constructability analysis, cost estimating, scheduling, value engineering, and site logistics planning. Construction is a risky business, and project managers and superintendents need to identify potential risks and select strategies for managing them. A risk register is a useful tool for identification and management of potential project risks. VE may be performed as a preconstruction service or may be conducted in response to a VE incentive provision in the construction contract. VE involves evaluating project systems and components to determine the most cost-effective materials for the project.

Some contracts contain provisions involving the use of a partnering process. It is a cooperative approach to project management that involves creating a team for project execution. Partnering begins with an initial workshop that involves selection of project goals, development of an effective communication system, and development of an issue resolution system. Since team members may be new to each other, a period of team building may be needed.

BIMs are used to share information, identify conflicts among design disciplines, simulate construction operations, and support off-site construction. If the project owner establishes a specific LEED goal for the project, the project manager and superintendent need to assist the designer in selecting specific credits to be achieved and collect documentation to support the certification application. Prevention of polluted storm water runoff is mandated in most jurisdictions and must be considered in preconstruction planning. Pull planning schedules and off-site construction techniques should be considered to minimize costs and maximize value.

## Review questions

1. What are five services that a project owner may require of the general contractor in a preconstruction services contract?
2. What is a risk register, and how is it used?
3. What is the difference between constructability analysis and value engineering?
4. What is meant by the term *life-cycle cost* when conducting VE studies?
5. Why was the concept of partnering developed?
6. What is a partnering charter, and how is it developed?
7. Whom do you recommend for participation in the initial partnering workshop?
8. Why is the partnering charter a critical document in developing and maintaining a partnering relationship?
9. What is an issue resolution system, and why is it a critical component of partnering?
10. How are building information models used for preconstruction planning?

11. How are building information models used for constructability analysis?
12. How are 4D building information models used for preconstruction planning?
13. What is the LEED building assessment system?
14. Why do some project owners pursue LEED certification?
15. What is a SWPPP, and why is it an important preconstruction planning tool?
16. What is pull planning, and why is it used for preconstruction planning?
17. What are the benefits of using off-site construction techniques?

## Exercises

1. You have been asked to conduct a VE study to determine the most cost-effective type of window system for a hospital project. What functions would you consider essential in your analysis, and what cost data would you consider in conducting the study?
2. You have been asked to organize a partnering workshop for a major hospital renovation project. Whom would you invite to the workshop? What specific issues should be addressed in the development of the partnering charter?
3. Who should participate in the partnering workshop for the NanoEngineering Building?
4. You are a project manager on a project seeking LEED gold certification. What measures would you take to minimize construction waste on the project?

# 6 Subcontracting

## Introduction

General contractors typically use subcontractors to execute most of the construction tasks involved in a project. Indeed, a typical general contractor subcontracts 80 percent to 90 percent of the project scope of work, while residential contractors may subcontract 100 percent of their scopes of work. Some industrial and highway general contractors may self-perform most of their scopes of work. Subcontractors, often referred to as *specialty contractors*, therefore, are important members of the general contractor's project delivery team and have a significant impact on the general contractor's success or failure. The relationship of the subcontractors to the other members of the project delivery team is illustrated in Chapter 1. Since subcontractors have such a great impact on the overall quality, cost, and schedule for a project, they must be selected carefully and managed efficiently. There must be mutual trust and respect between the superintendent and the subcontractors, because each can achieve success only by working cooperatively with the other. Consequently, project managers and superintendents find it advantageous to develop and nurture positive, enduring relationships with reliable subcontractors. Project managers and superintendents must treat subcontractors fairly to ensure the subcontractors remain financially viable as business enterprises. This will ensure that the subcontractors will be available for future projects.

General contractors use subcontractors (1) to reduce risk and (2) to provide access to specialized skilled craftsworkers and equipment. One of the major risks in contracting is accurately forecasting the amount and cost of labor required to complete a project. By subcontracting significant segments of work, the project manager can transfer much of the risk to subcontractors. When the project manager asks a subcontractor for a price to perform a specific scope of work, the subcontractor bears the risk of properly estimating the labor, material, and equipment costs. Craftspeople experienced in the many specialized trades required for major construction projects are expensive to hire and generally used on a project site only for limited periods of time. Unless general contractors have continuous need for such specialized labor, it is cost-prohibitive to continually employ them as a part of the contractor's own work forces.

Subcontracting is not without risk. The project manager and superintendent give up some control when working with subcontractors. The scope and terms of the subcontract define the responsibilities of each subcontractor. If some aspect of the work is inadvertently omitted, the project manager and superintendent are still responsible for ensuring the general contract requirements are achieved. Specialty contractors are required to perform only those tasks that are specifically stated in the subcontract documents. Consistent quality control may be more difficult with subcontractors, particularly the quality of workmanship. Owners expect to receive a quality project and hold the project manager accountable for the quality of all work whether performed by the general contractor's crews or by subcontractors. Subcontractor bankruptcy is another risky aspect of subcontracting that can be minimized by good prequalification procedures and timely payment for subcontract work. Scheduling subcontractor work often is more difficult than scheduling the general contractor's crews, because the subcontractor's workers may be committed to other projects. Safety procedures and practices among subcontractors may not be as effective as those used by the general contractor presenting an additional challenge to the superintendent.

The major subcontracting challenges to be faced are:

- defining well-understood, specific scopes of work;
- selecting qualified subcontractors;
- negotiating fair prices for the subcontracted scopes of work;
- scheduling and coordinating subcontractor work to ensure timely completion of the project;
- causing precedent subcontractors to perform their work on time and in accordance with contract requirements;
- resolving technical issues involving subcontractors' work;
- ensuring the quality of subcontractors' work;
- paying subcontractors on time for properly completed work; and
- developing enduring relationships with reliable subcontractors.

In this chapter, we will discuss subcontractor selection, acquisition, and management. Selection and acquisition of subcontractors are part of project start-up activities, as we will discuss in Chapter 8. Selecting quality subcontractors is essential, if the project manager and superintendent are to produce a quality project on time and within budget. Project managers must remember that poor subcontractor performance will reflect negatively on their professional reputations and their ability to secure future projects. Once the subcontractors have been selected, contract documents are executed documenting the scopes of work and the terms and conditions of the agreement. Subcontractor management is an integral part of project management. While the project superintendent manages the field performance of the subcontractors, the project manager manages all subcontract documentation and communication and is responsible for ensuring the subcontractors are treated fairly.

## Subcontract scopes of work

During the initial work breakdown, the project manager identifies the work items that are to be self-performed by the general contractor and those that are to be subcontracted. He or she develops a subcontract plan identifying which work items to include in each subcontract. This plan becomes the basis for creating the specific scope of work for each subcontract.

The subcontract scope of work must specifically state what is included and what is excluded. A well-defined scope of work is essential to ensuring prospective specialty contractors understand what is expected of them. Poorly defined scopes of work lead to conflicts and often result in cost escalation, time delays, and litigation. The subcontract must state clearly the exact scope of work to be performed and either include or make reference to all relevant drawings and specifications. This ensures there will be a clear understanding of the subcontractor's responsibilities.

The subcontract scope of work should contain:

- a description of the work to be performed,
- a list of specific scope items to be included, and
- a list of items that are to be excluded (if applicable).

The scope may include only construction services, or it may include both design and construction services. Construction services may include labor and equipment only or may include labor, materials, and equipment. Action words should be used in writing scopes of work to minimize misunderstandings. Words such as *provide* and *install* should be used to define clearly the subcontractor's responsibilities.

The following is an example of a scope of work involving labor and equipment only:

*Subcontractor agrees to provide all labor, supervision, equipment, supplies, and other items necessary or required to install the doors, frames, and hardware and perform all such work in accordance with the contract drawings, specifications, and any addenda contained in the general contract documents for the construction of the NanoEngineering Building. Exclusion: Doors, frames, and hardware materials will be provided by general contractor.*

The following is an example of a scope of work involving labor, materials, and equipment:

*Subcontractor agrees to provide all materials, labor, supervision, equipment, supplies, and other items necessary or required to perform all electrical work in accordance with the contract drawings, specifications, and any addenda contained in the general contract documents for the construction of the NanoEngineering Building. The intent is to provide and install a complete electrical system that meets local code requirements. Includes conduit required for installation of telephone cabling by others*

The following is an example of a scope of work involving design and construction:

*Subcontractor agrees to design the automatic fire protection sprinkler system for the NanoEngineering Building and provide all materials, labor, supervision, tools, equipment, supplies, and other items necessary or required to install the sprinkler system in accordance with the approved design. Designer must be licensed, and design must conform to local fire protection code. Subcontractor must obtain construction permit for designed system.*

Project managers must ensure that all work items to be subcontracted are included in a subcontract scope of work. Omitted items eventually become change orders to the subcontract providing the project manager unanticipated expenses. Project managers must also ensure that individual work items do not appear in more than one subcontract to avoid paying for them more than once. An audit of subcontract scopes of work is completed before specialty contractors are asked to submit cost proposals. This audit involves reviewing all work items identified in the work breakdown and ensuring that each work item selected for subcontractor performance is included in one, and only one, subcontract scope of work.

Now that we have concluded our discussion of the scope of work, let us examine subcontract documents.

## Subcontract documents

The subcontract describes the agreement between the general contractor and each specialty contractor used on the project. It usually is based on the requirements and specifications contained in the general contract with the project owner. Once the scope of work has been established, the specific terms and conditions are defined for each subcontract. These terms and conditions establish the operating procedures the project manager intends to use to manage each subcontract. Topics that should be addressed are:

- subcontract price,
- list of subcontract documents,
- commencement and progress of work,
- availability of temporary utilities,
- jobsite cleanup requirements,
- requirement for daily report,
- availability of lifting support and scaffolding,
- insurance,
- indemnification,
- bonds (if required),
- safety,
- disposal of hazardous waste,
- inspection and acceptance of completed work,
- change order procedures,
- claims procedures,
- payment procedures,
- warranty,
- dispute resolution, and
- termination or suspension.

Standard subcontract documents have been developed by the American Institute of Architects and ConsensusDocs. Project managers should use one of these documents, because they are used widely in the industry and understood by specialty contractors. The advantage of using copyright documents is that they have been tested in court and found legally sufficient. Project managers who choose not to use one of these standard subcontracts should consult with legal counsel when crafting subcontract language. Developing your own subcontracts is very risky and not recommended.

## Subcontractor prequalification

To ensure quality work, project managers should prequalify specialty contractors before asking them to submit a price for the work. Historically, too much emphasis has been placed on price and not enough on quality and experience. Project managers should select subcontractors based on their ability to provide the greatest overall value to the project. Proven performers should be selected based on their professional approach to producing quality work.

Potential subcontractors should be evaluated using a standard set of criteria, such as:

- experience and technical skills required;
- technical and supervisory competency of management and field supervisors;
- stability and financial strength;
- adequacy of specialized equipment and labor;
- past safety performance; and
- reputation for working cooperatively, proven history of responding to warranty calls, and past project performance.

One technique for gathering much of this information is to require prospective subcontractors to respond to a detailed questionnaire, such as the example illustrated in Box 6.1. Additional information should be acquired from other project managers employed by the general contractor, and personal interviews may be required before selecting the set of prequalified subcontractors invited to submit proposals. Owners and designers should be invited to participate in reviewing the prospective subcontractors' qualifications. Project managers should use only quality subcontractors as members of their project delivery teams. Specialty contractors reviewed during prequalification should be evaluated against a set of criteria similar to that listed in the preceding paragraph. Each firm should be ranked based on the project manager's assessment of its capability to provide quality service as a member of the project delivery team.

## Subcontractor selection

Selecting quality specialty contractors for each subcontract is essential for project success. Quality specialty contractors have good safety records, experienced craftspeople, specialized equipment, and adequate financial capability to complete the project to the desired standards without experiencing financial problems. Some project managers select subcontractors simply on price. This often leads to problems on the project with quality control or timely execution. The goal is to select best-value subcontractors. On some projects, the owner reserves the right to approve subcontractors. While qualification-based selection of subcontractors is highly recommended, some owners, particularly public agencies on cost-plus construction contracts, may require that an open bidding system be used for selection of subcontractors. Subcontracts are then awarded to the lowest responsive bidder.

The recommended strategy is to select the top-ranked five or six specialty contractors identified during prequalification and invite them either to submit a proposal or a bid (quotation) for the scope of work. Requests for proposal typically are used for design-build or cost-plus scopes of work. Invitations to bid are used for lump sum or unit price scopes of work. An

# Box 6.1    Subcontractor questionnaire

### NORTHWEST CONSTRUCTION COMPANY
*1242 First Avenue, Cascade, Washington 98202*
*(206) 239-1422*

### SUBCONTRACTOR QUESTIONNAIRE

1. **General Company Information:**

   Name:  *Pacific Mechanical*

   Address:  *270 SW 41st Street*

   City:  *Renton*                State:  *WA*              Zip:  *98057*

   Telephone:  *425-251-1562*                FAX:  *425-251-1580*

   Type of firm.  (check appropriate box)

   ☒    Sole owner.  Name:  *Joe Johnson*

   ☐  Partnership.  List names of partners.

   _____
   _____
   _____

   ☐  Corporation.

   President:_____

   Vice President:_____

   Number of years in business under present name: *20  years*

   Trade(s) normally performed by company: *HVAC, ductwork, plumbing, fire protection, and sometimes controls.*

   Number of trade and office personnel currently employed:

   Trade employees: *200*     Office employees: *12*

2. **Financial Information:**

   What is the maximum dollar value of work the company is capable of handling at one time?

   *$25,000,000*

   Attach last 2 years audited financial statements at end of questionnaire.

3. **Insurance Information:**

   What is the company's workers' compensation experience modification rate for the 3 most recent years?

   *2012: 1.2*                *2013: 0.8*                *2014: 0.6*

continued …

*Box 6.1* continued

How much insurance coverage does the company currently carry?

|  | Yes | No | Amount |
|---|---|---|---|
| General Liability | X |  | $10,000,000 |
| Automobile Liability | X |  | $10,000,000 |
| Workers' Compensation | X |  | As Required by State |

4. **Safety Information:**

Does the company have a written safety program?    *Yes*

Use your OSHA Form 200 to complete the following table.

|  | 2012 | 2013 | 2014 |
|---|---|---|---|
| Total Number of Workers' Compensation Claims | 5 | 4 | 3 |
| Number of Lost Time Workers' Comp. Claims | 5 | 4 | 2 |
| Number of Accident Liability Claims | 1 | 0 | 0 |
| Number of Fatalities | 0 | 0 | 0 |

5. **Project Information:**

List current, ongoing projects with approximate dollar value and estimated completion date.

| Project | Amount | Completion Date |
|---|---|---|
| Key Bank Building | $8,570,000 | November 19, 2015 |
| South Shore School | $5,650,000 | January 8, 2016 |
| Eastside Apartments | $2,500,000 | March 20, 2016 |
|  |  |  |

Has the company failed to complete any work assigned to it during the past 5 years?  *No*
If Yes, explain:

_____

Attach a list of projects completed in the last 3 years and a list of contractor references whom we may contact.

6. **Equipment Information:**

Attach a list of owned construction equipment with capacity, age, type, and attachments.

This questionnaire was completed by:

Name: *Joe Johnson*                                    Title: *Owner*

Signature: ***Joe Johnson***                        Date: *July 21, 2015*

example invitation to bid is shown in Box 6.2, and an example request for proposal is shown in Box 6.3. Most project managers solicit bids electronically using written solicitations to ensure that each prospective subcontractor understands the project requirements and submits bids on the same scope.

Subcontractor bids typically are received by the project manager electronically. Bids received from one subcontractor should not be disclosed to another. This practice is known as *bid shopping* and is considered unethical, as discussed in Chapter 1. Subcontractors expect project managers to treat their bids as confidential information until the subcontracts are awarded. Project managers who practice bid shopping may find subcontractors unwilling to bid on their projects.

Once the proposals or bids have been received, they are evaluated to select the specific specialty contractor for each subcontract. A subcontractor analysis form similar to the one shown in Chapter 3 (see Worksheet 3.4) can be used to assist in subcontractor selection. To ensure each subcontractor's cost proposal is realistic, the project manager should have previously developed an order-of-magnitude cost estimate for each subcontract. Unrealistically low or high bids generally indicate that the subcontractor:

- was uncertain regarding the exact scope of work,
- found ambiguities regarding some of the conditions of work,
- erred when preparing the quotation, or
- omitted items (for low bids).

A high bid also may indicate that the subcontractor is fully committed on other projects and not interested in the work. Any ambiguities regarding the scope of work should be resolved, and all subcontractors should be invited to submit new bids.

Subcontractor performance and payment bonds generally are required if the general contractor is required to provide bonds to the owner. If the owner does not require bonds, the general contractor may not require subcontractor bonds if the subcontractors are selected based on their technical competence, reputation, and financial condition. Sometimes, general contractors require bonds on subcontracts exceeding some value, such as $100,000. However, bonds generally are required by contractors if the contract with the owner requires subcontracts to be awarded using a public bid procedure. If bonds are required by the project manager as a part of the project risk management strategy, forms similar to those illustrated in Box 2.4 and Box 2.5 in Chapter 2 should be used. Bonds are not free. If required, premiums will be included in each subcontractor's bid or identified as an additive alternate. Bonds protect the project manager against subcontractor default, bankruptcy, and liens but do not guarantee successful subcontractor performance. Subcontractor bonds are similar to the general contractor bonds discussed in Chapter 2. The major difference is that the surety guarantees the subcontractor's performance to the general contractor rather than the general contractor's performance to the owner.

Once each specialty contractor is selected, the price is entered on the subcontract, and the agreement is signed by an executive from the general contractor and the specialty contractor. The procedures used are similar to those discussed in Chapter 2 for executing the contract between the owner and the general contractor. A ConsensusDocs 751 subcontract agreement for electrical work on the NanoEngineering Building is available on the companion website for this book.

## Box 6.2    Subcontract invitation to bid/subbid proposal

### NORTHWEST CONSTRUCTION COMPANY
1242 First Avenue, Cascade, Washington 98202
(206) 239-1422

### INVITATION TO BID/SUBBID PROPOSAL

To:    Pacific Mechanical                                              Date:  August 7, 2015
     Attention:  Joe Johnson
     270 SW 41$^{st}$ Street
     Renton, Washington 98057

     Project:                      NanoEngineering Building
     Project Address:      4010 Stevens Way, Seattle, Washington 98195

#### Part 1: Invitation

Northwest Construction Company invites Pacific Mechanical to submit a bid on or before 10:00 a.m. Pacific Daylight Time on September 11, 2015 for the following subcontract work:

Provide all materials, labor, supervision, equipment, supplies, and other items necessary or required to perform all mechanical work (plumbing, HVAC, and controls) in accordance with the contract drawings, specifications, and any addenda contained in the general contract documents for the construction of the NanoEngineering Building.  The intent is to provide and install a complete mechanical system that meets all code requirements.

The general contract plans, specification, general and special conditions and addenda are available online from Builders Exchange.  The project designer is Zimmer Gunsul Frasca and the mechanical engineering consultant is Acme Engineering.

#### Part 2: Bid Proposal

*Base Bid: $14,100,000*

*If this bid is accepted, the Subcontractor agrees to enter into a mutually acceptable subcontract and, if required, furnish performance and payment bonds from a surety licensed in the State of Washington and acceptable to the Contractor.  The cost of the bond is not included in this proposal, but will be added to the subcontract price at 1.5 " of the base bid.*

*This bid will remain in effect and not withdrawn by the Subcontractor for a period of 30 days or for the same period of time required by the contract documents for the Contractor in the prime bid, plus 15 days, whichever is longer.*

*Subcontractor:  Pacific Mechanical*

By:  **Joe Johnson**
*Title:  Owner*
*Date: September 11, 2015*

# Box 6.3   Subcontract request for proposal

## *NORTHWEST CONSTRUCTION COMPANY*
1242 First Avenue, Cascade, Washington 98202
(206) 239-1422

## REQUEST FOR PROPOSAL

To:   Puget Sound Fire Protection                    Date:  August 18, 2015
     Attention:  Thomas White
     1736 Detroit Street
     Seattle, Washington 98101

Project:                NanoEngineering Building
Project Address:        4010 Stevens Way, Seattle, Washington 98195

Northwest Construction Company is soliciting Design-Build proposals for the fire protection systems in the above named project.  Subcontract is to be awarded on a cost-reimbursable plus fee basis.  Proposals are due in three copies no later than November 4, 2015.  Selection will be dependent upon the overall subcontractor responsiveness with particular emphasis on the creative and detailed response to the "Proposed Systems", "Proposed Team", and "Proposed Budget" portions of this RFP.

Proposal Requirements:

1.  Subcontractor's Fee for the Work:
     State proposed fee as a percentage of the estimated cost of the work.
2.  Subcontractor Labor Cost:
     Provide a list of your proposed hourly labor costs for all direct and indirect reimbursable labor.
3.  Proposed Subcontractor Team and Management Plan:
     Provide a list of proposed project personnel including project manager, design engineer(s), and project superintendent.  Provide a comprehensive and concise project management plan detailing how you plan to complete the subcontract scope of work.  Describe how this team differentiates itself and what measures will be taken to provide competitive budgets, cost control, and innovation.
4.  Proposed System Narratives:
     Provide fire protection system design narrative including description of major components selected.
5.  Proposed System Detailed Budget:
     Provide a detailed fire protection  systems budget for project.

Sincerely,

*Ted Jones*
Ted Jones

## Subcontractor management

Once the subcontracts are awarded and construction begins, the project manager and the superintendent work together to schedule and coordinate the subcontractors' work to ensure the project is completed on time, within budget, and in conformance with contract requirements. Since most of the construction work is performed by subcontractors, efficient management of their work is critical to the project manager's ability to control costs and complete the project on time. Before allowing subcontractors to start work on the project, they should be required to provide a certificate of insurance. Coverage requirements should be stated in the subcontract. An example of a subcontractor insurance certificate is shown in Box 8.4 in Chapter 8.

The first challenge is to mold the subcontractors into a cohesive project delivery team. This requires an understanding of their concerns and proper work sequencing to ensure their success. It is essential that the superintendent establish a cooperative relationship with the subcontractors and their foremen by conducting frequent coordination meetings to discuss their concerns. Frequent, open communications coupled with mutual goals will help foster subcontractor relationships that are built on trust. The superintendent should require each subcontractor to submit a daily report of its activities. This provides a daily record of each subcontractor's progress and any obstacles encountered in performing the work. The requirement to submit daily reports must be included in the terms and conditions of each subcontract.

The success of any project is dependent on a viable schedule. The superintendent is responsible for coordinating start and completion dates with each subcontractor. They need adequate time to arrange their scheduled jobs, obtain equipment and materials, and schedule their crews. This means the superintendent must provide adequate notice to each subcontractor regarding the scheduled start date for their phase of the work. Timely notice is fundamental to building good relationships with subcontractors. The notice should also indicate the scheduled completion date to preclude interfering with the work of following subcontractors. This includes participating in pull planning work scheduling as discussed in Chapter 5. If different subcontractors will work concurrently on the project site, the superintendent must ensure that they are compatible and that each receives notice. All subcontractors must be aware of their safety responsibilities including those toward other subcontractors working on the site. The superintendent must ensure that the jobsite is ready for a subcontractor before scheduling him or her to start work. Requiring subcontractors to arrive on a site that is not ready causes a hardship on them and can result in lost time and potential increased project cost. The superintendent also is responsible for resolving any conflicts between subcontractors. Even though the work is being performed by a subcontractor, the general contractor is still responsible for the quality of the work performed, the schedule, and jobsite safety.

The superintendent must ensure subcontracted work conforms to quality requirements specified in the contract. Subcontractors must understand workmanship requirements before being allowed to proceed with the work. The superintendent should conduct preconstruction meetings with subcontractors before allowing them to start work on the project. A sample agenda for these meetings is shown in Box 6.4. The owner and designer should be invited to participate in all subcontractor preconstruction meetings. Often mock-ups, which are stand-alone samples of completed work, are required for exterior and interior finishes. This allows the superintendent and the designer to evaluate the work and establish a standard of workmanship. Project site cleaning policies also should be discussed at the preconstruction conference. Generally, all subcontractors should be required to clean their work areas when they have completed their scopes of work.

## Box 6.4   Subcontractor preconstruction meeting agenda

### Subcontractor Preconstruction Meeting Agenda

1. Introduction (Points of contact established)
2. Document status
   - Subcontract agreement signed
   - Insurance certificate received
   - Project specific safety plan received
3. Scope review
4. Work status
   - Submittal status
   - Permit status
   - Material status
5. Schedule review
6. Coordination and safety
7. Quality control
8. Field administration
   - Daily foreman's report
   - Jobsite maintenance and clean up
   - Jobsite meetings
   - Back charge notification
   - Requests for information
9. Contract administration
   - Pay requests – lien releases
   - Notification of changed conditions
   - Change order procedures
   - Close-out requirements

Superintendents need to actively oversee the work of all subcontractors to ensure that their work conforms to contractual requirements. Subcontractors must not be allowed to build over work that has been improperly done by a previous subcontractor. The superintendent should walk the project when each subcontractor finishes its portion of the work to identify any needed rework. An inspection report similar to the one illustrated in Box 6.5 should be used to document the results of each inspection. This is better than waiting until the end of the construction and trying to get several subcontractors to return and do rework. Early identification and correction of deficiencies is a key part of an active quality management program as will be discussed in Chapter 13.

## Box 6.5   Project inspection report

### NORTHWEST CONSTRUCTION COMPANY
1242 First Avenue, Cascade, Washington 98202
(206) 239-1422

### PROJECT INSPECTION REPORT

Project:  NanoEngineering Building                          Date:  March 16, 2016

Subcontractor:  Pacific Mechanical

Area of Work:  First floor mechanical rough-in

Participants:

  Tom Brown, Pacific Mechanical Foreman

  Harry Jones, Acme Engineering

  Jim Bates, Superintendent

Scope of Inspection:

  Drywall subcontractor is ready to install drywall on the first floor, so mechanical
rough-in

  needs to be inspected before being covered

| Items Requiring Correction | Date Corrected | Date Reinspected |
|---|---|---|
| 1. Hangers missing on water line to break room | 3/18/16 | 3/18/16–Bates |
| 2. Water supply to emergency eye wash facility in Room 115 missing | 3/18/16 | 3/18/16–Bates |

Subcontractors may have questions regarding some aspects of their scopes of work. A request-for-information procedure should be established to document subcontractor questions and the responses. Many questions can be answered by the superintendent or project engineer, but others may require information from the design team. Simple forms, such as the example illustrated in Chapter 10 (Box 10.4), should be used to expedite the processing of subcontractor inquiries. To ensure all subcontractor questions have been answered, a log similar to the one shown in Box 10.5 is maintained. The project manager should review the subcontractor request for information logs periodically to ensure timely responses are provided to subcontractors.

Situations may arise when the scope of work needs to be modified. All such modifications should be documented as subcontract change orders to identify clearly the changes in scope and the impact on the subcontract price. Sometimes changes affect multiple subcontractors. In these cases, all affected subcontracts must be changed appropriately. Prior to negotiating the cost of a change order, the project manager sends a change order request to the affected subcontractors. An example for the NanoEngineering Building project is shown in Box 6.6. When the initial contract for the NanoEngineering Building was awarded, there was insufficient funding to finish

---

## Box 6.6   Subcontractor change order request

### NORTHWEST CONSTRUCTION COMPANY
*1242 First Avenue, Cascade, Washington  98202*
*(206) 239-1422*

### SUBCONTRACTOR CHANGE ORDER REQUEST

Date: *July 14, 2016*

SCOR No.: *003*

Distribution:  *Richardson Electric*

Attached are the following documents:  *Revised drawings E3.03 and E4.03*

Description of Change: *Install complete electrical systems in both classrooms on first floor.*

Please prepare an estimate and schedule impact for the enclosed. Your response is required to be submitted to the above address, in detail, with all supporting estimating backup, no later than:

Date Due: *August 2, 2016*
Time Due: *Noon*

If a response is not received by the above date and time we will assume that the enclosed has no impact to your scheduled or estimated cost of work and will issue a change modification to your subcontract/purchase order incorporating these documents indicating no impact.

Submitted by: **Mary Peterson,** Project Engineer

File code: *9821/COP #003*

all spaces in the building including two classrooms on the first floor. Later, the University provided additional funding to enable the two rooms to be finished, requiring a change to the construction contract and an increase in the guaranteed maximum price. All change orders should be issued in written form, as illustrated in Box 6.7. A change order register such as the one shown in Box 6.8 should be maintained for each subcontract to document all changes.

---

## Box 6.7    Subcontract change order

### *NORTHWEST CONSTRUCTION COMPANY*
*1242 First Avenue, Cascade, Washington 98202*
*(206) 239-1422*

### SUBCONTRACT CHANGE ORDER

Change Order Number:  *003*

Project:  *NanoEngineering Building*

Subcontract:  *Richardson Electric*

We agree to make the following change(s) in the subcontract scope of work:

   *Install complete electrical systems in both classrooms on first floor*

Change(s) will affect the following plans and/or specifications:

   *Revised Drawings E3.03 and E4.03*

Subject to the following adjustment to the subcontract value:

   Cost of this Change:  *$85,150.00*

   Previous Subcontract Amount:  *$5,009,350.00*

   Revised Subcontract Amount:  *$5,094,500.00*

   Revised Completion Date:  *No change*

Contractor:  NORTHWEST CONSTRUCTION COMPANY

By:  **Ted Jones**  *Project Manager*          Date:  *August 8, 2016*
        Authorized Signature/Title

Subcontractor:  *Richardson Electric*

By:  **Rick Richardson**  *President*          Date:  *August 8, 2016*
        Authorized Signature/Title

# Box 6.8    Subcontract change order register

## NORTHWEST CONSTRUCTION COMPANY
*1242 First Avenue, Cascade, Washington 98202*
*(206) 239-1422*

### SUBCONTRACT CHANGE ORDER REGISTER

Project:  NanoEngineering Building

Subcontractor:  Richardson Electric

Date Awarded:  September 16, 2015

| Change Order Number | Date of Change Order | Scope of Change Order | Change Order Value | Subcontract Value |
|---|---|---|---|---|
| | | Original Subcontract Value | | $5,000,000.00 |
| 1 | 11/5/15 | Revised lab layout | $5,150.00 | $5,005,150.00 |
| 2 | 1/7/16 | Exterior lighting | $4,200.00 | $5,009,350.00 |
| 3 | 8/12/16 | Classroom electrical | $85,150.00 | $5,094,500.00 |
| --- | --------- | Changes reconciliation | ----------- | --------------- |
| | | | | |

Subcontractors submit requests for payment as their work progresses, at the end of each phase of their work, or according to the payment schedule provided in the subcontracts. An example subcontract payment request is shown in Box 6.9. Once the superintendent approves a payment request, it is included in the general contractor's payment request to the owner as discussed in Chapter 11. Most subcontracts contain a provision that the subcontractor will be paid for the work performed once the owner pays the general contractor. It states that the subcontractor relies on the credit of the owner, not the general contractor, for payment. Such contracts place significant financial burdens on subcontractors when owners fail to make timely progress payments to general contractors. There are other standard subcontracts that contain a provision that the general contractor will pay the subcontractor for work completed within a reasonable time, regardless of whether the owner has paid the general contractor for the work. These contracts place the total risk of owner nonpayment on the general contractor. Subcontractors should be paid timely to avoid causing them cash flow problems and to ensure their financial health. Before making final payment, the subcontractor should be required to submit a final lien release, similar to the sample shown in Box 6.10. Lien releases are also discussed in Chapter 17.

## Box 6.9   Subcontract payment request

# SUBCONTRACTOR APPLICATION FOR PAYMENT

To: Northwest Construction Company

From: Hi-Gloss Painting Company

Project:  NanoEngineering Building, 4010 Stevens Way, Seattle, WA 98195

Payment Request: #2

Period: August 1, 2016 to August 31, 2016

### Statement of Contract Account

| | | |
|---|---|---:|
| 1. | Original Contract Amount: | $590,000 |
| 2. | Approved Change Order Nos:  3: | $5,600 |
| 3. | Adjusted Contract Amount: | $595,600 |
| 4. | Value of Work Completed to Date: | $195,000 |
| 5. | Value of Approved Change Orders Completed: | $2,500 |
| 6. | Materials Stored on Site: | 0 |
| 7. | Value of Completed Work and Stored Materials: | $197,500 |
| 8. | Net Total to Date: | $197,500 |
| 9. | Less Amount Retained (5"): | ($9,875) |
| 10. | Total Less Retainage: | $187,625 |
| 11. | Total Previously Certified (Deduct): | $95,500 |
| 12. | Amount Due this Request: | $92,125 |

#### Certification of the Subcontractor

I certify that the work performed and the materials supplied to date, as shown above, represent the actual value of accomplishment under the terms of the subcontract (and all authorized changes thereto) between the undersigned and Northwest Construction Company relating to the above referenced project.

I certify that the payments have been made through the period covered by previous payments received from the contractor to all of my subcontractors and for all material and labor used in or in connection with the performance of this subcontract.  I further certify that I have complied with Federal, State, and local laws insofar as applicable to the performance of this subcontract.

Furthermore, in consideration of the payments received, and upon receipt of the amount of this request, the undersigned does hereby waive, release, and relinquish all claim or right of which the undersigned may have upon the premised above described except for claims or right of lien for contract and/or change order work performed to the extent that payment is being retained or will subsequently become due.

Subcontractor: Hi-Gloss Painting Company                    Date:  August 29, 2016

By:  *Mark Jackson*
        Authorized Signature

Title:  Owner

## Box 6.10   Final lien release

**FINAL LIEN RELEASE**
**UPON RECEIPT OF FINAL PAYMENT**

Upon receipt by the undersigned of a check from *Northwest Construction Company*
in the sum of *$25,200.00*   payable to *Hi-Gloss Painting Company*
and when the check has been properly endorsed and paid by the bank upon which it is
drawn, this document shall become effective to release any mechanic's lien rights the
undersigned has on the project of

*Painting work for NanoEngineering Building*

located at *Seattle, Washington.*
The undersigned has been paid in full for all labor, services, material and equipment
furnished to *Northwest Construction Company.*

Dated: *December 30, 2016*                     *Mark Jackson*

By:  *Mark Jackson, Owner*

## Summary

Subcontractors are essential members of the project manager's team. Typically, they perform most of the work on a construction project. During the initial work breakdown for a project, the project manager determines which work items are to be subcontracted and which of those work items to include in each subcontract. Based on this subcontract plan, the project manager crafts a specific scope of work for each subcontract. Once the scope of work has been developed, the project manager prepares the terms and conditions of the subcontract or selects a standard subcontract format.

Because subcontractors are so critical to project success, the project team should prequalify them before asking for price quotations. Subcontractors should be selected based on their ability to provide the greatest overall value as members of the project delivery team. Once the subcontractor has been selected, the price is entered on the subcontract, and the agreement is signed both by the project manager and the subcontractor.

Once the subcontracts are awarded and construction begins, the project manager and the superintendent work together to schedule and coordinate the subcontractors' work to ensure the project is completed on time, within budget, and in conformance with contract requirements. The success of the project is dependent on a viable schedule that provides adequate notice to all subcontractors regarding the scheduled start times for their phases of work. Subcontract management includes ensuring quality performance of work, responding to requests for information, issuing change orders when needed, and promptly paying subcontractors for accepted work.

## Review questions

1. Why do general contractors use subcontractors rather than performing all the work on a project?
2. What risks do general contractors incur by using subcontractors?
3. Why might a project manager require performance and payment bonds from a subcontractor?
4. Why is a termination or suspension provision included in most subcontracts?
5. Why is subcontractor bid shopping considered unethical behavior?
6. Why do subcontractors need reasonable notice regarding the scheduled start times for their phases of work? Why does the superintendent not determine the exact start times for each subcontractor at the beginning of the project?
7. What are mock-ups, and how are they used for project quality control?
8. How are subcontractors' requests for payment processed on a typical project?
9. Why is a subcontractor's ability to finance his or her cash flow requirements on a project of concern to the project manager?
10. What is the difference between a request for quotation and a request for proposal? What type of subcontract is awarded using a request for quotation? What type is awarded using a request for proposal?

## Exercises

1. Write a clear scope of work for a subcontract for the site work subcontract for the NanoEngineering Building project.
2. What are five criteria that you suggest be used to prequalify site work contractors for the NanoEngineering Building project?
3. What basis do you suggest the project manager use to select the electrical subcontractors for the NanoEngineering Building project?

# 7

# Material management

## Introduction

Building materials are needed to assemble the project. As discussed in Chapter 6, some materials are furnished by the subcontractors, while materials for self-performed work are procured by the general contractor. Materials requiring special manufacturing or long-lead time also may be procured by the GC for installation by subcontractors. This gives the contractor control of the cost of these items and the ability to influence their delivery times. However, the GC assumes responsibility for any subcontractor impacts caused by late delivery of contractor-procured materials. Materials should be ordered early in the construction process to minimize the risk of price inflation and to ensure construction activities are not delayed because of a lack of needed materials.

Material deliveries should be phased so materials arrive on the project site near the time they are needed to support construction activities. Most project sites are not large enough to store all needed materials on site at the same time, and just-in-time delivery of materials avoids congesting the construction site with storage of materials. In addition, stored materials need to be protected against theft, damage, and weather. Damaged materials do not meet contract specifications and must be replaced at the contractor's expense. Limiting the length of on-site storage minimizes the cost of providing adequate protection.

In lump sum contracts, owners may choose not to pay contractors for materials stored on the project site but wait until the materials are incorporated into the project. Early delivery of materials on such projects may adversely impact the contractor's cash flow because material suppliers may request payment before the owner pays the contractor for the materials. On some cost-plus contracts, like the one used in our example case study project, the cost of materials stored on site is reimbursable.

Major materials are those that require submittal approvals and have a significant lead time or delivery time. Examples of these include structural steel, reinforcing steel, specialized mechanical or electrical equipment, and some interior finish items. Such materials typically require designer approval before being ordered. The normal procedure is to first select the supplier; issue a PO; receive submittals; obtain approval of the submittals from the design team; receive the materials on-site; and store the materials until installed. Submittals include shop drawings, product data sheets, material samples, and others as described below. Shop drawings are drawings or diagrams prepared by suppliers to illustrate their products. Product data sheets are used to illustrate performance characteristics of materials described in shop drawings or submitted as verification that the materials meet contract specifications. The submittal process is discussed below. A member of the design team verifies that the submittals, in whatever format, correctly interpret contract requirements. Material dimensions are the responsibility of the general contractor.

Structural steel is a material common to many commercial and civil construction projects. Structural steel typically is fabricated by a supplier and erected by a subcontractor. First, the fabrication shop drawings are approved by the GC and the structural engineer. Then the supplier cuts the steel elements to the correct dimensions and drills the holes for bolted connections. The steel erection subcontractor submits shop drawing showing the erection sequence. Once these have been approved, the subcontractor can begin erecting the steel. The typical material procurement process is illustrated in Figure 7.1.

Materials are ordered using supply contracts called *purchase orders* (POs). Suppliers provide invoices with the materials at time of delivery. Contractors make payment to suppliers using the invoices as reference documents. Some suppliers offer material discounts if the invoices are paid within a specified time after material delivery. Project managers should analyze their cash flow capabilities and may take advantage of these discounts to increase their profit margins. This may mean paying suppliers before being paid by the owner. A unique credit account and cost code generally is established with each supplier for each project, and expenditure limits are established by company policy.

## Supplier selection

Just as the selection of quality subcontractors is critical to the project manager and superintendent's ability to deliver a quality project, so is the selection of reliable, quality suppliers. Material requirements for a project are determined from the contract plans and specifications. The plans provide the quantitative requirements, while the specifications provide the qualitative requirements. Some suppliers might only sell materials, while others may manufacture and sell the materials. An example of the former is a lumberyard that sells lumber, masonry materials, and hardware. An example of the latter is the concrete supplier who mixes the concrete and delivers it to the site ready for placement. Suppliers typically are asked for price quotations before POs are executed. Similar to our discussion in the last chapter with subcontractors, suppliers should be selected based on the quality of their products, pricing, and their ability to deliver the materials to meet construction schedule requirements.

## Purchase orders

POs are contracts for the manufacture and/or sale of materials and equipment. There are two different types of POs used on construction projects. The first is a long-form PO that is used for major material purchases, typically prepared by the project manager or the construction firm's

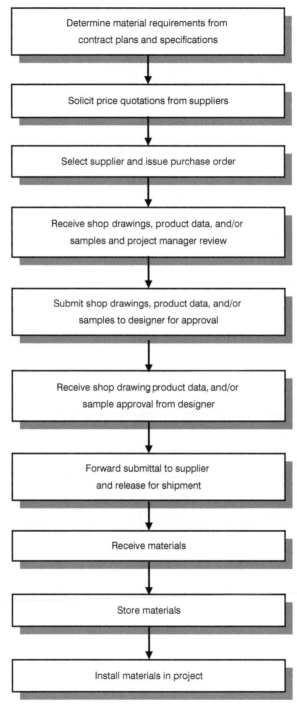

**Figure 7.1**    Material procurement process

purchasing department. Box 7.1 is the cover sheet of a sample PO from the NanoEngineering Building. There are often several additional pages of terms and conditions attached. The complete example of this PO is shown on the companion website. The second type of PO is a short-form PO, illustrated in Box 7.2, typically used in the field by superintendents to order materials from local suppliers. The superintendent may be assisted by a PE when preparing short-form POs. When preparing a PO, the construction team needs to provide a complete description, quantities

## Box 7.1    Long form purchase order

### MATERIAL PURCHASE ORDER CONTRACT

## Northwest Construction Company

1242 First Avenue
Cascade, Washington 98202
(206) 239-1422

CONTRACT NO.  *9821/03.32*                          ORDER NO.: *101*

FOR: *NanoEngineering Building*                     QUOTATION

TO: *Steel Fabrication, Inc.*                        TERMS: *30 days*
    *Post Office Box 113*
    *Seattle, WA 99801*                          DATE: *June1, 2015*

SHIP TO:  *Jobsite:*                           VIA: *Your trucks*
          *1105 15th Ave NE*
          *Seattle, WA 99801*          REQUIRED DELIVERY DATE: *10/07/15*

| Item No. | Quantity | Description | Code or Equip. No. | Unit Price | Amount |
|---|---|---|---|---|---|
| | | *Supply all reinforcement steel for installation by Northwest Construction Company* | 03-20-00 | | |
| | | | | | |
| | | | | | |
| | | *Excludes Installation* | | *lump sum* | $58,000.00 |

This contract is not assignable by Seller.
This Purchase Order Contract is subject to the terms and provisions printed on the reverse hereof.

SELLER:                                    BUYER:
By *Henry Smith*                         Northwest Construction Company
Title: *Project Manager*                 By *Ted Jones*

# Box 7.2    Short form purchase order

## PURCHASE ORDER

Northwest Construction Company
1242 First Avenue
Cascade, Washington 98202
(206) 239-1422

To:    *Seattle Lumber*
       *1141 Northlake Ave.*
       *Seattle, WA 98101*

Attention:    *James S. Black*
Phone No.:    *206.447.1642*

P.O. No.:    *1042*
Date:        *3/1/16*
Job Name:    *NanoEngineering*

Job Address:    *1105 15th Ave. NE*
                *Seattle, WA 98101*

| When Required: *3 Days* | | | | Ship Via: *Your Truck* | | |
|---|---|---|---|---|---|---|
| No. | Quantity | Unit | Description | | Unit Price | Amount |
| 1 | *100* | *Shts* | *¾" MDO Parapet Plywood* | | *$35/Sht* | *$3,500* |
| 2 | *2* | *Box* | *8d Nails* | | *$75/Box* | *$150* |
| 3 | *10* | *Tubes* | *Carpenter's Glue* | | *$3.5/Tube* | *$35* |
| 4 | | | | | | |
| | | | | | TOTAL | *$3,685* |

Purchase Order Instructions:

1. Accept orders only from duly authorized Buyer and obtain a purchase order number at time of purchase.
2. All packages or pieces must show this purchase order number.
3. Do not back order or substitute without prior approval.

1. Render separate invoices for each shipment.
2. Invoices received without purchase order number will be considered incorrect.

| Requested By: | Approved By: | Purchased By: |
|---|---|---|
| *Jim Bates* | *Jim Bates* | *Mary Peterson* |
| Superintendent | Superintendent | Project Engineer |

required, unit prices, and total cost for each item. Extracts from the specifications may be attached to clarify material or performance requirements. An example of a material specification for concrete reinforcement steel is also shown on the companion website.

When drafting POs, the project manager or PE must decide where the contractor desires to assume ownership of the materials. The alternatives are either the shipping point (supplier's warehouse) or the jobsite. Most materials will be ordered FOB (free on board, also known by some as freight on board) jobsite, which means the price quoted by the supplier includes the transportation to the jobsite. In this case, the supplier selects the transporting carrier. In some instances, it might be advantageous to order materials FOB shipping point, or FOB warehouse, which means the price quoted by the supplier does not include transportation costs. In this case, the contractor selects the carrier and pays the transportation cost but, in doing so, assumes the risks associated with damage during shipping and delivery delays.

Project managers should maintain a PO log, similar to the one shown in Worksheet 7.1, to manage their procurement activities. Some materials, such as ready-mix concrete, may be purchased on an open or blanket PO. The project superintendent simply contacts the supplier providing the mix number, quantity, and required time of delivery. An open PO is similar to a unit price contract. The contractor and supplier agree to a unit price, but the exact quantity to be used is not stated. The cost invoiced to the contractor is the quantity delivered multiplied by the unit price. A superintendent may request deliveries on different dates using the same PO. Open POs often contain a not-to-exceed quantity or cost and have an expiration date.

## Submittal management

The project manager's and superintendent's primary responsibilities are to complete the project on time, safely, within budget, and to specified quality requirements. The submittal process is a key part of the overall quality management program for the project. In this chapter, we will discuss submittals and the submittal process for material suppliers. Submittal concepts and processes are the same for subcontractors who supply their own material. Additional discussion of quality management is contained in Chapter 13.

A submittal is a document or product turned in by the construction team to verify that what they plan to purchase, fabricate, deliver, and ultimately install is in fact what the design team intended by their drawings and specifications. It serves as one last check and validity of design. Submittal requirements for a project are contained in the specifications of the contract. Project managers should look upon submittals as a first step in quality control and, therefore, as a tool to be used to complete a successful project. The procedures for processing submittals generally are in Divisions 00 and 01 of the contract specifications. Section 00-72-00.4.03 of the specifications contains submittal procedures for the NanoEngineering Building. This section of the special conditions contains specific instructions regarding submittal preparation, copies to be provided, and the review process.

## Types of submittals

A submittal can be any of the following:

*   coordination drawings,
*   cut sheets of product data,
*   shop, fabrication, or installation drawings,
*   samples and color charts, or
*   mockups.

Regardless of the type, submittals are one of the final design steps and also one of the first quality control steps. The construction drawings prepared by the designer, although adequate

Worksheet 7.1   Purchase order log

## PURCHASE ORDER LOG

Project No.: 9821          Project Name: NanoEngineering Building          Project Manager: Ted Jones

| P.O. No. | Supplier | Materials Delivered | Date Sent | Date Materials Due | Date Materials Received | Date Supplier Paid |
|---|---|---|---|---|---|---|
| 101 | Steel Fabricators | Reinforcement Steel | 6/1/15 | 10/7/15 | | |
| | | | | | | |
| 145 | Seattle Lumber | Surveying and flagging stakes | 8/15/15 | 8/17/15 | 8/17/15 | 9/01/15 |
| 146 | Seattle Lumber | 2x4s and plywood | 8/20/15 | 8/20/15 | 8/21/15 | 9/01/15 |
| 147 | Arctic Mechanical | Temporary drain pipe | 8/20/15 | 8/25/15 | 8/24/15 | 9/01/15 |
| 148 | Seattle Lumber | Silt fence | 9/02/15 | 9/09/15 | | |
| 149 | University Redi-Mix | 1 load quarry spalls | | | | |
| | | | | | | |

for cost estimating and general construction, do not show sufficient detail to be suitable for fabrication and production of many required construction products. Manufacture of required materials often requires that the contract drawings be amplified by detailed shop drawings that supplement, enlarge, or clarify the project design. These descriptive shop drawings are prepared by the manufacturers or fabricators and provided to the purchasing organization, either the GC or a subcontractor. These shop drawings are reviewed by the GC's PE to ensure conformance with contract drawings and specifications before being forwarded to the designer. Submittal cover sheets, similar to the transmittals discussed in Chapter 10, are used to transmit shop drawings to the designer.

Design-build subcontractors, such as mechanical, electrical, and plumbing (MEP), generate submittals as well, but technically they are the also the designer so their submittals should be forwarded to the GC stamped 'approved' by themselves. The client and architect may still want to review these submittals in regards to process equipment or fixtures. Additional design-build subcontractors include firms such as shoring and curtain-wall, and their design efforts should also be submitted to the design team for disposition. Often the subcontractors will utilize CAD and BIM and place multi-disciplines' work on the same drawings called *coordination drawings*. These are intended to resolve conflicts on paper or, in this case, on the computer screen, prior to materials arriving on the jobsite and saving all firms substantial costs in re-fabrication and re-work.

Other submittals are used to demonstrate that materials selected for the project conform to contract requirements. These may be performance requirements; descriptive requirements, such as color of carpeting or wall covering; or proprietary requirements. SDS sheets for all materials are an essential element of a project safety plan, and these documents often accompany a supplier's submittal cut sheet package. Again, submittal cover sheets are used to transmit manufacturer's technical data and product samples to the designer.

## Review and approval

There have been recent changes in the *approval* that is received back from the designer. It is difficult today to find a submittal stamp or any other correspondence from a designer, with the word *approved* on it. The word *approved* often is replaced with *reviewed, conforms,* or *no exceptions taken.* This also occurs on many design professionals' field reports wherein *inspected* has been replaced with noncommittal words such as *visited* or *witnessed* or *reviewed* or *observed.* The wording used on approval stamps has been selected so that the reviewer does not assume any of the legal responsibility of the party seeking approval.

Some designers will require the GC to stamp a submittal *approved* prior to the designer's *review.* Most contractors also have noncommittal stamps that are similar to those of many designers. Some general contractors use a stamp that indicates *received* and *forwarded.* Many general contractors are now asking the subcontractor and supplier to stamp a submittal *approved* prior to submitting it, therefore passing some of the responsibility to them. To some extent, this is a good practice. It requires the submittal originator and the GC to at least read the submittal and not just pass it through. Some designers return submittals without any stamp, signature, or comment. Regardless of what the stamp says, it is recommended that project managers submit, submit early, submit everything, and submit often. A good argument can be made that if the reinforcing steel shop drawings were turned in to the structural engineer and they were returned without any marks or comments, they have been accepted. After all, they were not rejected. Unfortunately many owners get caught in the middle with neither party taking responsibility.

At a minimum, the construction team must submit everything that the specifications require. If the specifications do not specifically require that the fire extinguishers be submitted, yet the specifications clearly call out the color of the extinguishers as "purple," the project manager has reason to believe that this is an error. If an error is assumed and red extinguishers ordered, the project manager may be surprised. Conversely, if a specified unique color is ordered and the owner and designer are surprised during the final inspection, turnover may be affected. The project manager should be proactive. Even though a fire extinguisher submittal was not required, the project manager should still forward one. In the long run, the designers will appreciate this sort of early notification and validity check. Many subcontractors and suppliers fight the submittal process. It is in their best interest to participate fully. The more documents that are received back from an owner or architect with their approval signature and stamps on them, regardless of format, the better off the subcontractors and suppliers are.

## Submittal processing

Submittals also allow the project manager to identify some of the hidden errors and exceptions the subcontractors and suppliers have taken in their bids. Although they bid it "per plans and specifications," and although the subcontract clearly reinforced this, the project manager does not want to be surprised 3 months later when the wood doors that the supplier has delivered are really cherry veneer and not solid cherry. Although the project manager may be contractually correct, this does not help get the building turned over, the owner moved in, and retention released.

Sometimes project managers do not want the designer to 100 percent complete the design. Sometimes it is in the construction team's best interest to have some of the choices of field routing, material selections, and construction applications left up to the contractor. This is referred to as "means and methods," an area most owners and architects do not want to step into, and those who do, shouldn't have. Examples would include the choice of concrete forming systems or a structural steel joint that, if bolted in lieu of welded, would be safer and assist with the installation.

The submittal process also is one of the early checks of the validity of the construction schedule. If the toilet accessory cut sheets are late being submitted by the supplier, a good case could be made that the delivery will also be late. The timing of submittals, as well as deliveries and construction installation, should be noted on all of the POs and subcontracts.

Owners and architects cannot always make up their minds on all possible product choices prior to awarding the contract. An example would be plastic laminate counter top colors. If a product is specified, such as "ACME Manufacturing" with the note, "color to be selected later by the owner," this is adequate to get competitive pricing and issue contracts.

Submittal planning involves the development of a schedule of submittals that is given to the architect for review and an expediting log to manage the submittal process with subcontractors and suppliers. A flow chart of the submittal planning process is shown in Figure 7.2. Article 3.16 of the ConsensusDocs 500 contract agreement for the NanoEngineering Building available on the companion website contains the requirement for the general contractor's management of submittals. Shortly after receiving award of the contract, the project manager should review each

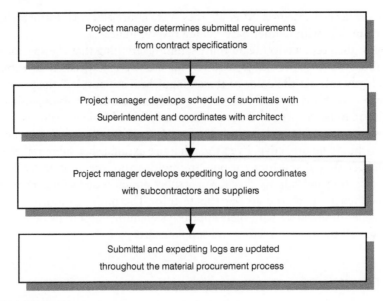

**Figure 7.2**  Submittal planning

specification section for submittal requirements. Language should be included in each subcontract and PO regarding quantity and timing of submittals. The construction schedule is then reviewed with the superintendent, and submittals are scheduled. A preliminary submittal list or expediting log with all potential submittals is prepared. Some items may be grouped or dropped at a later date. An example is shown in Worksheet 7.2. A letter is sent to each subcontractor and supplier with the expediting log attached asking for their concurrence or exceptions to the items that will be submitted as well as the date materials are required.

The process for managing submittals was illustrated in Figure 7.1. After a submittal is received, it should be reviewed by the PE with respect to the requirements in the specifications and the drawings. The project manager does not want to be in a position of explaining why the floor covering contractor submitted vinyl tile when slate was clearly required. Vendors have been known to attempt to substitute a cheaper, non-specified product. The shop drawings should be reviewed for requested dimensions. The submittal should be stamped as discussed earlier. Each submittal is assigned a number and entered on a submittal log. This log is updated and discussed at the owner/ architect/contractor coordination meeting as well as at subcontractor coordination meetings. This is different from the expediting log. Expediting issues and problems with subcontractors should be kept out of meetings with the owner. Discussion should be limited to submittal materials and documents that the designers have received. Some will assign a submittal number for every item prior to processing any submittals. These numbers can be tied into the specification section numbers. For example, reinforcing steel may be simply submittal number 5 for the entire project or may be number 03-20-00-1 for the first submittal in reinforcing steel specification 03-20-00. Regardless of the numbering system utilized, the important point to remember is to number each submittal individually. This allows for efficient tracking and future reference. An abbreviated submittal log is illustrated in Worksheet 7.3. After submittals are received and verified by the project engineer, a submittal cover sheet is completed. This is similar to a transmittal, except it

# Worksheet 7.2   Expediting log

## EXPEDITING LOG

Project No.: 9821     Project Name: NanoEngineering Building     Project Manager: Ted Jones

| No. | Description | Supplier | Committed Shop Dwg Submittal | Actual Shop Dwg Submittal | Fabricate/ Delivery Duration | Scheduled Delivery Date | Actual Delivery Date | Responsible Person |
|-----|-------------|----------|------------------------------|---------------------------|------------------------------|-------------------------|----------------------|--------------------|
| 1 | Import sample | Earth Enterprises | 8/20/15 | 8/20/15 | NA | 9/6/15 | 9/6/15 | Peterson |
| 2 | Pea gravel | Earth Enterprises | 8/20/15 | 8/20/15 | NA | 9/14/15 | 9/13/15 | Peterson |
| 3 | Silt fence | Seattle Lumber | 8/20/15 | 8/21/15 | NA | 9/9/15 | 9/10/15 | Peterson |
| 4 | Perforated pipe | Seattle Lumber | 8/20/15 | 8/28/15 | 2 days | 9/20/15 | 9/16/15 | Peterson |
| 5 | Footing rebar | Steel Fabrication | 8/1/15 | 8/7/15 | 2 weeks | 10/7/15 | | Peterson |
| 6 | Embed steel | Steel Fabrication | 9/6/15 | 10/7/15 | 2 weeks | 9/22/15 | 9/19/15 | Peterson |
| 7 | Concrete mix | University Redi-Mix | 9/6/015 | 9/15/15 | NA | 10/20/15 | | Peterson |
| 8 | SOG wire mesh | Steel Fabrication | 9/1/15 | | 1 week | 9/24/15 | 9/24/15 | Peterson |
| 9 | Asphalt specification | ATB Asphalt | | | NA | 9/14/16 | | Peterson |
| 10 | Column rebar | Steel Fabrication | 9/1/15 | | 2 weeks | 10/25/15 | | Peterson |
| | | | | | | | | |

# Worksheet 7.3  Submittal log

## SUBMITTAL LOG

Project No.: 9821          Project Name: NanoEngineering Building          Project Manager: Ted Jones

| No. | Originator | Sub's No. | Specification Section | Description | Date Submitted | Date Return Requested | Date Returned | Disposition* |
|---|---|---|---|---|---|---|---|---|
| 1 | Earth Enterprises | 1 | Civil drawings | Import sample | 8/20/15 | 8/27/15 | 8/26/15 | A |
| 2 | Earth Enterprises | 2 | Civil drawings | Pea gravel pipe bedding | 8/20/15 | 8/27/15 | 8/26/15 | A |
| 3 | Northwest Construction | NA | Civil drawings | Silt fence data sheet | 8/21/15 | 8/22/15 | 8/23/15 | A |
| 4 | Northwest Construction | NA | Civil drawings | Perforated pipe sample | 8/28/15 | 9/15/15 | 9/10/15 | AN |
| 5 | Steel Fabrication | 1 | 032000 | Footing rebar shop drawings | 8/7/15 | 8/14/15 | 8/15/15 | AN |
| 6 | Steel Fabrication | 2 | Structural drawings | Steel embed shop drawings | 10/7/15 | 10/14/15 | | |
| | | | | | | | | |

*A — approved, AN — approved as noted, RR — revise and resubmit, and R — rejected

is used exclusively for submittals. Each construction firm will have its own format. The architect may have a required form and may have even included it in the specifications. A submittal cover sheet is illustrated in Box 7.3. The approval stamps discussed above should be applied to the actual submittal documents, not the cover sheet.

## Box 7.3   Submittal cover sheet

### NORTHWEST CONSTRUCTION COMPANY
*1242 First Avenue, Cascade, Washington 98202*
*(206) 239-1422*

### SUBMITTAL

| | |
|---|---|
| Project: *NanoEngineering Building* | Date: *August 7, 2015* |
| Area/System: *Footing Rebar* | Submittal Number: *5* |
| To: *ZGF Architects* | First Submittal? *Y* |
| Address: *522 West 10th Street* | Re-Submittal? *NA* |
| *Seattle, WA 99801* | Previous Number(s): *NA* |
| Attention: *Norm Riley* | |

Required Response Date: *August 14, 2015*

Subcontractor/Supplier providing submittal: *Steel Fabrication*   Originator's Number: *1*

| Specification Section | Description | Drawing No. | Action Taken |
|---|---|---|---|
| 032000 | *Two sheets of spot footing reinforcement steel shop drawings* | *S1 & S2 Prints* | *Checked* |

Remarks by Contractor: *Please review and approve. Keep one set of prints for record. Return balance for copying and distribution.*

Submitted by: *Mary Peterson, Project Engineer*

Space below for Architect/Engineer:

Reply: *Approved,* Approved as noted, *Revise & Resubmit, Rejected*

Signed: *Norm Riley, ZGF*                    Date: *August 15, 2015*

Contractor's Job No./File Code: 9821/032000

Submittals may be processed and approved by the architect, the owner, the city, consultants, sub-consultant designers, or a combination of the above. They may be reviewed by several people, either concurrently or sequentially. The project manager or PE should set up the submittal log to be able to document the entire review process. After receipt of the processed submittal from the reviewing party, the project manager needs to review it for disposition and log it accordingly. If changes have been made to the products by the designer, the project manager needs to make note for potential change orders. Submittals should be reviewed both by the PE and by the superintendent when they are first received from the subcontractor or supplier and again after review by the design team.

The submittal should be sent back to the originating party, who will then take action. Other contractors and suppliers may need to review the final submittals. For example, a steel erector will need to see the steel fabricator's drawings. It is the supplier's or subcontractor's responsibility to notify the project manager if any change of scope has arisen due to comments received. The designer and owner may use the submittal process not only as a vehicle to finish the design; they may use it to change the design. If a change of scope is included, for example changing all of the #4 reinforcement steel to #5, a formal change document such as a construction change directive (CCD) should have accompanied the submittal return. CCDs and the change order process are discussed in Chapter 15.

## Scheduling material deliveries, including just-in-time deliveries

Project managers should initiate material procurement early in the construction process to ensure materials are available on site when needed by the construction workers. Special manufactured items and items that are shipped long distances must be ordered early. The project manager wants to order materials early to lock in prices to avoid price inflation. Adequate time must be allowed for the submittal and review of shop drawings, product data, and product samples. Some materials that are typically ordered, expedited, and submitted early include:

- select import backfill,
- shoring,
- concrete reinforcement steel,
- concrete mix designs,
- elevators,
- windows, and
- hollow-metal doorframes.

The project manager or PE may prepare long-form POs for major materials, or they may be prepared by the construction firm's purchasing department. In either case, the project manager or PE must establish the qualitative requirements, quantities, and required delivery dates. Supplier submittals must be reviewed by the project manager or PE to ensure that materials are adequately described in the submitted documents and materials conform to contract requirements. Poor submittals often are rejected by the designer and must be resubmitted, which may result in delayed material deliveries and late project completion. Required delivery dates are established based on material need dates derived from the project schedule. Material deliveries should be scheduled so materials arrive on site at the time they are needed for installation on the project. This just-in-time approach to material delivery minimizes the need for material storage on the project site. Materials must be ordered early enough to allow normal transportation modes to be used. Expedited transportation, such as air freight, is costly and may not have been considered when developing a bid or proposing a guaranteed maximum price.

Occasionally the project manager may order materials delivered earlier than they can be received at the jobsite and stored nearby. This might be done to lock in a favorable price, ensure material is available for installation when needed, or allow installers (and designers and owners) to visibly inspect the materials or equipment to make sure that adequate space has been allowed or service connections are sized properly. In this case, the project manager will attempt to have these materials paid for by the client before actual jobsite delivery. See Chapter 11 for additional discussion on the pay request process.

## Material management at project site

Construction labor productivity is greatly influenced by the organization of the project site and the flow of equipment, labor, and materials through the site. Site planning is discussed in the next chapter, but here we will examine some issues related to material management on the project site. Material storage sites should be selected where they have the least impact on the efficient construction of the project. This means they typically are not adjacent to the project being constructed. In determining the size and location of material storage areas, the project manager and the superintendent must anticipate the material requirements of the entire project, including those provided by the subcontractors. Materials should be stored as near as possible to the location of installation without adversely impacting the productivity of the workforce. Storage areas should be organized so materials are moved only once on the project site, either from the truck or storage location to the place of installation. Double-handling material cuts into profit margins. Materials should be stored so they are accessible when needed. Material delivery routes should be selected to avoid impacting construction operations. Materials procured by the GC should be inspected upon delivery to ensure that the correct items and the correct quantities were delivered. This is an important part of an active quality management program that is discussed in Chapter 13. Materials should be properly protected from the weather until installed. Pallets or dunnage typically are placed under stored materials to keep them off the ground. Materials damaged in storage often are not accepted by the client and must be replaced at the contractor's cost.

## Summary

Effective material management is essential to the timely completion of the construction project. Construction materials must be procured in time to meet project schedule requirements. Most of them may be provided by subcontractors, but some typically are procured by the general contractor. The project manager decides which materials the contractor will procure and initiates the procurement process. Suppliers are selected based on their ability to provide quality materials at a reasonable price and in the time frames needed to meet schedule requirements. Long-form POs are used by the project manager to purchase major materials. Short-form POs are used in the field to order materials from local suppliers.

Submittals are documents or products that are submitted to the designer for review. They represent the final phase of the design process and are key components of the project manager's quality management program. Submittals include shop drawings, manufacturer's technical data, and product samples. They may be prepared by subcontractors, suppliers, or the project manager's technical staff. Because of their potential impact on the construction schedule, all submittal requirements should be identified at the beginning of the project and tracked using a submittal log. The submittal log is used for managing the submittal process with the design team, and an expediting log is used for managing the submittal process with subcontractors and suppliers. An effectively managed submittal program is a necessary tool to achieve a successful construction project.

Required material delivery dates must meet project schedule requirements. Just-in-time deliveries are often necessary due to limited site storage and may be the most cost-efficient approach. Effective material management on the project site is essential for efficient construction operations. Materials should be handled only once on the project site and stored near the point of installation but in an area that does not adversely impact productivity on the project.

## Review questions

1. Why is effective material management an essential project management function?
2. What type of construction materials would a GC typically purchase on a long-form PO on a project involving the construction of a four-story structural steel office building?
3. Why are submittals (shop drawings, product data, or product samples) required from suppliers? Who reviews these submittals?
4. Why are submittal approvals obtained before the supplier ships the required materials?
5. How would you determine the required delivery date for the stone needed to construct the exterior walls for the NanoEngineering Building?
6. What criteria should a project manager use when selecting a material supplier for a project?
7. How does the organization of a jobsite affect material management efficiency on a construction project?
8. What is a PO? What are its main components? Why should a project manager keep a PO log?
9. Why might a project manager choose to pay suppliers for materials received before being paid by the project owner?
10. What criteria should the project manager use in selecting material storage areas on a construction work site?
11. Who is responsible for preparing submittals?
12. Who is responsible for approving submittals?
13. What is the difference between a contract drawing and a shop drawing?
14. List five items that would require a submittal on a wood-frame multi-family construction project.
15. Why would a contractor not want the design team to completely design an item or system of work? List three examples.
16. How is a submittal considered part of the quality control process?
17. How is a submittal considered part of the design process?
18. List two different methods of numbering a submittal.
19. When is the submittal log reviewed?

## Exercises

1.  Prepare a long-form PO for the procurement of a ton of #8 reinforcing steel. Contact a local reinforcement steel supplier for a unit price.
2.  Prepare a short-form PO for the procurement of 50 pounds of 16d nails, 30 joist hangers, and 1,100 board feet of 2x8 pressure-treated joist for an outdoor deck. Contact a local lumber store for associated unit prices. What other materials would your foreman need to build this deck?
3.  Prepare a submittal coversheet for the hollow metal doorframes for the NanoEngineering Building. Use the project schedule and other documents included on the companion website to assist in identifying approximate dates and lead times.
4.  Prepare a submittal coversheet for an item on the NanoEngineering Building project that may not normally require a submittal. Select an item that may involve some uncertainty. Explain why you selected the item.

# CHAPTER

# 8 Project start-up

## Introduction

Project start-up is a sub-set of preconstruction planning as was discussed in Chapter 5. Once the project manager has been notified that the project has been won, he or she must plan project start-up activities. Efficient project start-up activities are dependent on a good start-up plan. Start-up activities involve establishment and organization of the project office and staff, construction site layout, and mobilization of the jobsite. The project manager selects the project management team and involves them in start-up planning. Contract files, project office administrative procedures, and correspondence management systems are also established. These tools will be discussed in Chapter 9. Award of subcontracts, procurement of materials, and actual mobilization must wait until the notice to proceed (NTP) is received, but the planning should be completed prior to its issuance, so implementation can begin immediately upon receipt of the notice. The actual mobilization onto the jobsite is the responsibility of the project superintendent.

Both the project manager and the superintendent participate in the preconstruction conference. This meeting is conducted by either the owner or the designer to introduce project participants and to discuss project issues and management procedures. The special conditions of the NanoEngineering Building specifications discusses the preconstruction conference, processes, and includes the preconstruction services agreement, as was discussed in Chapter 5. The NTP authorizes the contractor to start work on the project and often is issued at the end of the preconstruction conference. An example NTP is shown in Box 8.1.

Once the NTP has been received, the project manager should buy out the project by awarding subcontracts and ordering needed materials and equipment, especially those long-lead items. The NTP often is provided at the preconstruction conference along with verification of owner financing and needed permits. Project-specific quality control and safety plans need to be finalized and, if required by contract, submitted to the owner or owner's representative. Quality control planning is discussed in Chapter 13, and safety planning is discussed in Chapter 14. A schedule

## Box 8.1    Notice to proceed

# NOTICE TO PROCEED

**To:**    *Northwest Construction Company*          **Date:** *March 25, 2015*
*1242 First Avenue*
*Cascade, Washington 98202*

### Project Name:  *NanoEngineering Building*

You are hereby notified to commence WORK in accordance with subject contract executed *March 24, 2015* on *March 25, 2015* and you are to complete the WORK within *690* consecutive calendar days thereafter.  The date for substantial completion of all WORK is therefore *January 15, 2017*.

By:      ***Jeffrey Jackson***
Title:    *Director, Capital Projects*
Owner:          *University of Washington*

### ACCEPTANCE OF NOTICE

Receipt of the above NOTICE TO PROCEED is hereby acknowledged by *Northwest Construction Company* this the *26th* day of *March, 2015*.

By:      ***Sam Peters***
Title:    *Vice President*
Contractor: *Northwest Construction Company*

of values and a schedule of submittals also are prepared and submitted to the owner or owner's representative for approval. The schedule of values is discussed in Chapter 11, and submittals were discussed in Chapter 7.

## Jobsite planning

Efficient jobsite organization is essential for a productive construction project. The jobsite layout affects the cost of material handling, labor, and the use of major equipment by the general contractor and the subcontractors. A well-organized site has a positive effect on the productivity of the entire project workforce and their safety. The jobsite layout plan should identify locations for temporary facilities, material movement, material storage, and material handling equipment. The proper choice of tower crane types and locations, as was the case on the NanoEngineering Building, is essential to a productive plan. Although developed by the superintendent, it should consider the needs and requirements of all subcontractors working on site. The site plan should:

- show existing conditions, such as adjacent buildings, utility lines, and streets;
- indicate the planned locations for all temporary facilities such as fences and gates, trailers, temporary utilities, sanitary facilities, erosion control, and drainage; and
- indicate areas for material handling, material storage, equipment storage, worker and visitor parking, and material handling equipment such as a tower crane.

In developing the jobsite layout plan, the superintendent should consider site constraints, equipment constraints, job site productivity, material handling, and safety. Some additional factors that should be considered in developing the jobsite plan are to:

- eliminate any bottlenecks to equipment movement on site;
- locate material storage areas near where the materials are to be used;
- locate staging areas where they are accessible to material handling equipment, such as cranes;
- place project office near the main entrance for visitor control;
- provide turnaround areas for haul vehicles; and
- minimize traffic impact on adjacent streets.

The jobsite layout plan for the NanoEngineering Building is shown on the text companion website. In addition to serving as a productivity tool, the jobsite layout plan is a great proposal/ interview/marketing tool. It shows the owner that the contractor has thought through the project—a personal touch. It may make a difference on a close award decision with a negotiated project.

Two separate project gates may be needed, if both union and non-union workers are to be employed on the project site. These are used to separate the two groups of workers. These gates will apply to the contractor's workforce, those of all subcontractors, and suppliers making material deliveries. One gate is used by union firms and their employees; the other by non-union firms and their employees. In the event of a labor dispute, only the union gate may be picketed by disputing union members. This practice minimizes the potential for costly job stoppages caused by labor disputes and is allowed under the National Labor Relations Act. To avoid confusion, each gate must be labeled conspicuously with a sign similar to the one illustrated in Box 8.2, and each subcontractor and supplier must be notified in writing which gate to use.

## Mobilization

The actual mobilization and physical move onto the jobsite is the superintendent's responsibility. The project office is established, and temporary utilities (electricity, water, sewer, and telephone) are installed. Site access is secured with adequate fencing, and gates are installed to provide needed site access for craftsmen and material delivery. Erosion control measures required by the storm

**Box 8.2    Subcontractor gate sign**

> ### THIS GATE RESERVED EXCLUSIVELY FOR THE EMPLOYEES, AGENTS, SUBCONTRACTORS AND SUPPLIERS OF:
>
> Richardson Electric, Inc.
>
> Acme Roofing Company
>
> Smith Fencing Company
>
> Hi Gloss Painting Contractors
>
> **All other persons must use the gate located on 15<sup>th</sup> Avenue on the east side of the project**

water pollution prevention plan discussed in Chapter 5 are installed to keep soil-laden water from leaving the project site and entering streams or wetlands. A project sign is installed at a visible location on the project site. Mobilization occurs soon after receipt of the NTP to demonstrate the contractor's commitment to starting the project.

Mobilization for the project manager involves organizing the project team and setting up the project files, control systems, and project logs. A start-up checklist similar to the one shown in Box 8.3 will help the project manager to ensure that all needed tasks are completed. The submittal and expediting logs were discussed in Chapter 7, the RFI log is in Chapter 10, the schedule of values is in Chapter 11, and the change order proposal log is in Chapter 15. The filing system and project files are important tools of the project manager. They must be organized in a logical sequence and clearly identified. The test of a good filing system is the ease with which needed records can be retrieved. The project manager also ensures that all contractually required initial documentation is submitted to the owner. At a minimum, this will include a schedule of values and a project schedule. It also may include a subcontracting plan, a quality control plan, and a project specific safety plan.

A start-up log such as the one shown in Worksheet 8.1 is a useful tool for managing start-up activities relating to subcontractors and suppliers. The project manager should require a certificate of insurance from each subcontractor before allowing them to start work and provide one to the owner ensuring that all parties working on the project have required coverage. A typical certificate of insurance is shown in Box 8.4.

## Box 8.3   Start-up checklist

**NORTHWEST CONSTRUCTION COMPANY**
*1242 First Avenue, Cascade, Washington 98202*
(206) 239-1422

**PROJECT START-UP CHECKLIST**

Project: *NanoEngineering Building*

Project Manager: *Ted Jones*

|  | Scheduled Completion | Date Completed |
|---|---|---|
| Cost codes set up: | 3/26 | 3/25 |
| Work packages set up: | 5/1 | |
| Buyout complete: | 7/1 | |
| Draft contract complete: | 2/11 | 2/11 |
| Certificate of insurance submitted to owner: | 3/20 | 3/22 |
| Subcontracting plan submitted to owner: | 4/1 | |
| Safety plan submitted to owner: | 4/1 | 3/28 |
| Quality control plan submitted to owner: | 4/1 | |
| Subcontracts drafted: | 7/1 | |
| Subcontracts awarded: | 8/1 | |
| Subcontractor insurance certificates received : | 9/1 | |
| Subcontractor safety manuals received: | 9/1 | |
| Subcontractor preconstruction meeting conducted: | 8/15 | |
| Building permit in hand: | 3/2 | 3/2 |
| Construction schedule submitted to owner: | 2/15 | 2/14 |
| Schedule of values submitted to owner: | 3/20 | 3/22 |
| Telephone list completed: | 8/1 | |
| Change order proposal log established: | 7/1 | |
| Purchase order log established: | 7/1 | |
| RFI log established: | 7/1 | |
| Expediting log established: | 7/1 | |

# Worksheet 8.1   Start-up log

## PROJECT START-UP LOG

**Project Number:** 9821          **Project Name:** NanoEngineering Building          **Project Manager:** Ted Jones

| | | | | | | | |
|---|---|---|---|---|---|---|---|
| **Sub/PO Code:** | 00.01.15 | 31.00.00 | 32.12.16 | 03.30.00 | 03.20.00 | 22.00.00 | 26.00.00 |
| **Category:** | Temp Fence | Mass Ex | Asphalt | Rebar Fab | Redi-Mix | Plumbing | Electrical |
| **Sub/Supplier:** | Smith Fence | Earth Ent | ATB Asph | Steel Fab | Westside | Rainy City | Richardson |
| **Letter of Intent:** | NA | NA | 4/1/15 | 4/1/15 | NA | 4/1/15 | 4/1/15 |
| **Rough Sub/PO:** | NA | 2/15/15 | 4/15/15 | 5/15/15 | NA | 5/1/15 | 4/15/15 |
| **Final Sub/PO:** | 3/25/15 | 3/25/15 | 5/1/15 | 5/15/15 | 5/15/15 | 5/1/15 | 4/15/15 |
| **Sub/PO Returned:** | 3/26/15 | 5/1/15 | 5/5/15 | | 5/30/15 | | 5/1/15 |
| **Insurance Rec'd:** | 3/26/15 | 5/1/15 | | NA | NA | NA | 5/1/15 |
| **Insurance Expires:** | 12/21/15 | 10/31/15 | | NA | NA | | 1/31/16 |
| **Safety Plan Rec'd:** | NA | 5/1/15 | NA | NA | NA | | |

## Box 8.4   Certificate of insurance

### CERTIFICATE OF INSURANCE

| **Producer** (Insurance Broker)<br><br>*ACME Insurance Co.* | THIS CERTIFICATE IS ISSUED AS A MATTER OF INFORMATION ONLY AND CONFERS NO RIGHTS UPON THE CERTIFICATE HOLDER. THIS CERTIFICATE DOES NOT AMMEND, EXTEND OR ALTER THE COVERAGE AFFORDED BY THE POLICIES BELOW. | | |
|---|---|---|---|
| | **COMPANIES AFFORDING COVERAGE** | | |
| **Insured** (Contractor)<br><br>*Reliable Glaziers* | | INSURANCE COMPANY | BEST RATING |
| | Company Letter A | *National Insurance* | *AAA* |
| | Company Letter B | *National Insurance* | *AAA* |
| | Company Letter C | *National Insurance* | *AAA* |
| | Company Letter D | | |

### COVERAGES

THIS IS TO CERTIFY THAT THE INSURANCE POLICIES LISTED BELOW HAVE BEEN ISSUED TO THE INSURED NAMED ABOVE FOR THE POLICY PERIOD INDICATED. NOTHWITHSTANDING ANY REQUIREMENT, TERM, OR CONDITION OF ANY CONTRACT OR OTHER DOCUMENT WITH RESPECT TO WHICH THIS CERTIFICATE MAY BE ISSUED OR MAY PERTAIN, THE INSURANCE AFFORDED BY THE POLICIES DESCRIBED BELOW IS SUBJECT TO ALL THE TERMS, EXCLUSIONS, AND CONDITIONS OF SUCH POLICIES. LIMITS MAY HAVE BEEN REDUCED BY PAID CLAIMS.

| Co. Ltr. | Type of Insurance | Policy Number | Policy Effective Date (M/D/Y) | Policy Expiration Date (M/D/Y) | Limits | |
|---|---|---|---|---|---|---|
| A | **General Liability**<br>X Commercial General    Liability<br>–<br>– | 27ABC649 | 10/01/15 | 09/30/16 | General Aggregate<br>Products/Comp Ops Agg<br>Pers. & Adv. Injury<br>Each Occurrence<br>Fire Damage (any one fire)<br>Med. Expenses (any one person) | $1,000,000<br>$1,000,000<br>$1,000,000<br>$1,000,000<br>$50,000<br>$10,000 |
| B | **Automobile Liability**<br>X Any Automobile<br>– | 27ABC659 | 10/01/15 | 09/30/16 | Combined Single Limit<br>Bodily Injury (per person)<br>Bodily Injury (per accident)<br>Property Damage | $1,000,000<br>$500,000<br>$1,000,000<br>$ 500,000 |
| C | **Excess Liability**<br>X Umbrella Form<br>– | 27ABC789 | 10/01/15 | 09/30/16 | Each Occurrence<br>Aggregate | $5,000,000<br>$7,000,000 |
| D | Workers' Compensation | | | | Each Accident<br>Disease Policy Limit<br>Disease (each employee) | $<br>$<br>$ |

### PROJECT NAME OR ALL OPERATIONS:

*Northwest Construction Co., University of Washington, and ZGF Architecture* are additional insured per attached endorsement. The liability insurance referred to in this certificate is primary and non-contributory to any insurance carried per attached endorsement.

| **CERTIFICATE HOLDER** | **CANCELLATION** |
|---|---|
| *Northwest Construction Company* | SHOULD ANY OF THE ABOVE DESCRIBED POLICIES BE CANCELED BEFORE THE EXPIRATION DATE LISTED, THE ISSUING COMPANY WILL PROVIDE 7 DAYS WRITTEN NOTICE TO CERTIFICATE HOLDER NAMED TO LEFT. |

**AUTHORIZED REPRESENTATIVE**

*Richard S. Sommerset*

The various types of insurance used in construction were discussed in Chapter 2. The general contractor typically is required to demonstrate proof of liability insurance and required to name the owner as an additional insured. This protects the owner and the general contractor from financial loss due to damage to third-party property or injury to third parties. Builder's risk insurance protects the project under construction by covering the cost of damage due to fire or severe storms. This insurance is purchased either by the owner or by the general contractor. Whoever purchases the insurance needs to ensure the other party is named in the policy as an additional named insured. Worker's compensation insurance is no-fault insurance that covers the cost due to injury or death of an employee on the project site. Such insurance is required by law and may be obtained from an insurance company or, if the state has a monopoly, must be purchased from the appropriate state fund. Worker's compensation insurance is also discussed in more detail in Chapter 14. Insurance requirements are typically contained in the supplemental or special conditions of the contract.

## Initial submittals

The technical specifications of the construction contract may require specific shop drawings, product data sheets, or samples to be submitted to the designers for review and approval. These are called *submittals* and are used by design professionals to ensure that correct materials are used on the project. They were discussed as part of the material-purchasing process in Chapter 7. The supplemental or general conditions of the contract typically discuss submittal requirements. Shop drawings are produced by the general contractor, subcontractors, or suppliers to show construction details not shown on the contract drawings. Shop drawings typically are required for systems such as reinforcing steel, precast concrete, structural steel, duct work, electrical equipment, fire protection systems, millwork, casework, metal doors, curtain walls, and manufactured roofs. Product data sheet submittals, or "cut sheets," generally are manufacturers' product information describing model or type, performance characteristics, and physical characteristics. Samples are physical parts of specified products. The specification section 00-72-00.4.03 contains the procedures used on the NanoEngineering Building; a brief excerpt follows:

> Shop drawings are provided by the Contractor showing in detail the proposed fabrication and assembly of structure elements; and the installation of materials and equipment. Submittals include, but are not limited to drawings, diagrams, layouts, schematics, descriptive literature, illustrations, schedules, performance and test data, samples and similar materials furnished by the Contractor...

An abbreviated example submittal specification requirement is the following for the carpet floor covering system, section 09-68-13.1.3, for the NanoEngineering Building:

> Shop Drawings—include layout of joints; Product Data describing sizes, patterns, colors and methods of installation; Samples; Manufacturer's installation instructions, and Maintenance Data.

See the companion website for the complete specification section for carpet and other systems referenced in the body of this text.

To manage the submittal review and approval process, the project manager often is required by contract to prepare a schedule of all submittals required by the contract specifications. Submittals need to be turned in and approved before the products are installed on the project. Submittals for products needed for early phases of the project should be submitted as a part of project start-up. This is to provide time for the designers to approve them and for the materials to be delivered to the project site without delaying scheduled activities. Some early submittals typically include:

- samples of earthwork backfill material,
- concrete batch mixes,
- reinforcement steel (rebar) shop drawings,
- structural steel shop drawings,
- hollow metal door frames,
- elevators, and
- long-lead mechanical and electrical equipment.

Proper management of the submittal process is critical to quality control and schedule adherence of the project. Late submission of submittals often leads to late delivery of materials and late completion of scheduled elements of work.

## Project buyout

The process of project buyout involves awarding subcontracts and procuring materials being furnished by the general contractor and comparing the actual costs to the budgeted costs. The purpose is to determine the status of the project with respect to the contractor's budget. A log such as the one illustrated in Box 8.5 can be used to buy out the project. The major unknown, which is not reflected on the buyout log, is the labor cost for self-performed work. This must be monitored weekly to track actual expenditures versus budgeted costs, as will be discussed in Chapter 12. Most project managers will choose to use electronic spread sheet software, such as Microsoft Excel, to compare buyout costs with budgeted costs.

As was discussed in Chapter 6, subcontractors should be selected based upon the best-value they contribute to the project team. Selection solely on price often leads to problems with quality and timely execution. Prior to being invited to submit proposals, the specialty contractors should be prequalified. A subcontractor proposal analysis form, similar to the one illustrated in Chapter 3, can be used to assist in subcontractor selection. The project manager wants to ensure that the intended scope of work is included in the price quotation, and he or she uses a detailed spreadsheet to compare all proposals for each scope of work. Prior to selecting the specialty contractor for each subcontract, the project manager and superintendent should conduct pre-award meetings with the best-value firms. The best-value firms are those who propose reasonable prices, have experienced craftspeople, are in sound financial condition, have excellent safety records, and have reputations for high-quality work. The project manager and superintendent meet with one specialty contractor at a time to:

- review the drawings and specifications to ensure the correct scope of work was considered in preparing the proposal;
- review any exclusions listed on the specialty contractor's proposal;
- discuss any questions submitted with the proposal; and
- discuss the size and receive commitments for the contractor's proposed workforce and schedule.

**Box 8.5  Buyout log**

## PROJECT BUYOUT LOG

Project Number: _9821_   Project Name: _NanoEngineering Building_   Project Manager: _Ted Jones_

| Code | Category | Budget | Buyout | Variance | Date | Sub/Supplier | Comments |
|------|----------|--------|--------|----------|------|--------------|----------|
| 01.40.00 | Temp fence | $15,000 | $15,000 | $0 | 3/25 | Smith Fence | Rental allowance |
| 31.00.00 | Mass ex | $1,949,000 | $1,999,000 | $50,000 | 5/1 | Earth Enter. | Added spoils |
| 32.12.16 | Asphalt | $26,000 | $25,050 | -$950 | 5/1 | ATB Asphalt | Picked up striping |
| 32.17.23 | Pavement stripe | $5,000 | $0 | -$5,000 | | 5/1 | w/ATB |
| 03.30.00 | Rebar supply | $250,000 | $250,000 | $0 | 5/15 | Steel Fab | |
| 03.20.00 | Redi-mix supply | $1,500,000 | $1,650,000 | $150,000 | 5/30 | Westside | Admixtures |
| 09.30.00 | Carpet | $1,800,000 | $954,000 | -$806,000 | 2/15 | Soft Touch | As Bid |
| 09.65.43 | Vinyl | w/carpet | $750,000 | $750,000 | 2/15 | United | As Bid |
| 22.00.00 | Plumbing | $2,750,000 | | | | Rainy City | Negotiating |
| 26.00.00 | Electrical | $4,900,000 | $5,000,000 | $100,000 | 5/1 | Richardson | Temp Power |
| 01.50.00 | Temp Power | $150,000 | $0 | -$150,000 | 5/1 | Richardson | was w/GCs |

Based on the results of the pre-award meetings and annotated subcontractor proposal analysis forms, subcontractors are selected for all scopes of work that will not be performed by the general contractor's workforce. Each subcontract value is then entered in the project buyout log. Buyout will also be discussed with cost control in Chapter 12.

## Three-week schedules

While contract schedules, such as those discussed in Chapter 4, are developed for overall project control, short-interval schedules generally are used by the superintendent to manage the day-to-day activities on the project. These short-interval schedules are developed each week throughout the duration of the project. Three-week schedules typically provide sufficient information for managing the project. The initial 3-week schedule must be prepared during project start-up to schedule the contractor's and subcontractors' workforces to minimize interference and smoothly plan the flow of work. The initial 3-week schedule developed by the superintendent for the NanoEngineering Building was included with Chapter 4.

In addition to the jobsite layout plan, scheduling, and participating with subcontractor selection, some other important start-up activities for the superintendent include preconstruction meetings. One is held with the city to review traffic flow and work hours. A separate meeting will be held with the city if there are temporary shoring requirements. If this were a union project, another preconstruction meeting is chaired by the superintendent with all the local union representatives to resolve any jurisdictional disputes before construction starts.

## Summary

The project manager and the superintendent are responsible for planning project start-up and mobilizing the jobsite. This involves planning the organization of the jobsite, organizing the project office, and physically mobilizing on site. The organization of the jobsite has a significant impact on the productivity of the entire workforce. The jobsite layout plan should identify locations for temporary facilities, material movement, material storage, and material handling equipment. The project manager mobilizes by organizing the project team and establishing the project files, control systems, and project logs. Initial submittals must be prepared and submitted for review so construction will not be delayed for lack of materials. The project manager buys out the project by awarding subcontracts and purchase orders. An initial short-interval schedule is prepared by the superintendent to schedule the trades needed to initiate construction.

## Review questions

1. Why does the project manager not initiate project start-up activities prior to receipt of the NTP?
2. What are three factors that should be considered when planning the organization of a jobsite?
3. What type of information should be shown on a jobsite layout plan?
4. Why are two entrance gates sometimes required?
5. What are the superintendent's responsibilities during jobsite mobilization?

6. What are the project manager's responsibilities during jobsite mobilization?
7. What are certificates of insurance, and why are they used?
8. What is workers' compensation insurance?
9. What are submittals, and how do they impact the construction schedule?
10. What is meant by project buyout?

## Exercises

1. Draw a site plan for a construction site involving the construction of an L-shaped three-story hotel. Include all of the items listed under the first three bullets in the section on jobsite planning above.
2. Prepare a start-up checklist for a major subcontractor, such as the electrician on the NanoEngineering Building.
3. Prepare a (1) short-interval schedule, (2) action plan, or (3) start-up checklist for the superintendent on the NanoEngineering Building in preparation of the concrete subcontractor's work. Coordinate this plan with the detailed project schedule and the specifications. Refer to other topics within the text such as QC and safety and insurance within this plan.

# 9 Project documentation

## Introduction

The physical move onto the jobsite and beginning construction work are the responsibility of the superintendent. The project manager mobilizes in a different fashion. Although on larger projects, he or she also will move into the job office or trailer. The project manager's mobilization responsibilities are focused primarily on starting the paper flow. This includes setting up the document systems and logs, employing subcontractors, and staffing the site management team. While previous chapters have discussed some of the project manager's contractual start-up responsibilities, this chapter covers many of his or her document and record-keeping responsibilities. In this chapter, we will discuss manual and electronic documents and record keeping. Project safety documentation and record keeping is discussed in Chapter 14.

## Project files

The project filing system and the files are important tools for the project manager. Files, properly prepared, are used in managing the project. They are not just storage locations. Documents must be filed in a logical sequence to facilitate easy retrieval. The key to a good filing system is, first, that it has a logical organization and, second, that documents are filed properly. All the information on a particular subject or involving a specific subcontractor should be placed in the appropriate file. This includes documents such as submittals, letters, change orders, and telephone conversations. The file should be kept in chronological order with the most current information on top.

The filing system and the file codes to be used should be set up early in the job. It is not necessary to set up job files during the bidding process but, as soon as the project manager is notified of a successful bid, the file system should be established. The files may be set up by the project manager or by a designee, such as the project engineer. The project clerk or project engineer (PE) may be the one to maintain and file project information during the course of the job, but the project manager should develop and take responsibility for the organization and accuracy of the filing system.

The file coding system should be either alpha or numeric. Often subcontractors' codes correlate with their CSI sections of the contract specifications. The important lesson here is:

- that some coding system be used, and
- that the coding system be used consistently throughout the construction firm.

It is recommended that the codes assigned to estimate line items match those in the filing system, those on the contracts, and also those on the invoices. Because files are used as a resource, they need to be located where the project manager can have ready access. If the project management team is working on a medium to large project, say greater than $50 million, the team likely is physically located at the site, and the files should be at the site. One exception to this is the original contract documents. These should be located in fireproof files and safely secured in the construction firm's corporate office.

The job files not only serve as an active project management tool during construction, they provide a permanent written record of the project. If the project is unfortunate enough to end up in a legal dispute, this is one of the first places the attorneys will begin their review. After the project's completion, the files should be kept safe for an established period of time. Sometimes this duration is dictated by local statute, other times by the contract terms, and still others by the construction firm's corporate philosophy. A common period is 3 to 7 years after the project is completed and all claims are resolved.

Of course, many project files today are kept electronically, but when there is a dispute, these files will all be printed and made available to the parties involved. Paper copies are also consistently used as backup to change orders, claims, and unfortunately, lawsuits.

## Superintendent's daily report

The daily report, or job diary, is an important construction management document. Some construction professionals refer to the diary as a daily journal, daily report, or daily log. The name is not as important as the fact that the document is used properly. The job diary is one of the most important project management tools and jobsite records. It is viewed as a contemporary record of each day's events. Its uses include:

- legal record of daily jobsite activity,
- change order backup documentation,
- back charge and claim documentation, and
- historical record of schedule progress.

An example of a job diary for the NanoEngineering Building is shown in Box 9.1. There are a variety of forms that are utilized throughout the industry. Each construction firm should select a standard format to use on all projects. The important items to be included with any diary include:

- superintendent's name,
- job name and reference number,
- date and day the diary is filled out,

## Box 9.1   Job diary

### NORTHWEST CONSTRUCTION COMPANY
*1242 First Avenue, Cascade, Washington  98202*
*(206) 239-1422*

**Daily Job Diary**

Project:            *NanoEngineering Building*            Job Number: *9821*
Superintendent:   *Jim Bates*     Date: *August 12, 2015*     Day: *Wednesday*
Today's Weather:   *Sunny, 76 degrees, Slight SW Wind at 5 MPH*

**Activites Completed:**
*Towercrane foundation has been completed and passed inspection. The TC rental firm along with the erection firm met onsite to discuss erection procedures for next weekend. Foundation work within the building pad continues. Electrical and Plumbing under-slab rough-in continues – still pending inspections.*

**Problems Encountered:**
*The University will not allow early delivery of towercrane materials as summer school is still in session. We felt that due to the smaller enrollment we would be able to mobilize on Thursday and Friday. Also we are going to need to pay for an official City of Seattle Police Officer for traffic control, our flagger will not suffice.*

**Materials and Equipment Received Today:**
*Replacement forklift (#2), additional plumbing and conduit, last footing rebar shipment*

**Rental Equipment Returned:**
*Forklift #1 which had ongoing hydraulics problems*

**On-Site Labor/Crafts/Subcontractors/Hours Worked:**
Carpenters:       *3 men  and 1 woman at 8 hours/day*
Laborers:         *4 all day, clean up, concrete prep*
Ironworkers:      *1 at 10 hours, stayed late to prep for pour*
CM Finishers:         *1 guy doing some patch and sack, 4 hours*
Masons:           _____
Electricians:     *2 men at 8 hours, 1 apprentice at 5 hours*
Drywallers:       _____
Plumbers:         *2 men, including a non-working foreman, at 8 hours/day*
HVAC Tinners:     _____
Painters:         _____
Other:            *TC inspector Don Smith for 3 hours*

- weather,
- manpower for both the general contractor and subcontractors,
- routing or distribution of copies,
- hindrance to normal progress,
- deliveries received,
- visitors to the site,
- equipment on site,
- accidents, and
- inspections.

The project superintendent should fill out the diary. Often this is delegated to the assistant superintendent, foreman, or field engineer. However, because of the historical importance of the diary, it should be prepared by the project superintendent, who is responsible for all field activities. The diary should be completed daily, at the end of the day. Postponing completing the diary until the next morning, or doing five of them at the end of the week, dilutes their accuracy. The diary should be shared with those who have a need for the information, such as the project manager and the office-in-charge of the project. On negotiated projects, the owner also may request copies. Some superintendents have been known to write one diary on Monday and make four photocopies, just changing the date for the remaining 4 days of the week. Others have been known to prepare two diaries, one for distribution externally, say to the client or architect, and the other for an internal record. These practices completely eliminate the validity and purpose of this important construction management tool.

Ideally, the diary should be written out by hand in ink. This helps support authenticity and originality. Many superintendents are, of course, now preparing their diaries on their computers. To some extent, this also negates the originality of the document. Printing and signing by hand in ink is beneficial. There are two schools of thought regarding loose-sheet or bound diaries. If they are loose, they can be copied, distributed, and filed easily, but loose sheets may also get lost. If they are bound and consecutively numbered, there is less chance of a diary rewritten to reflect a desired viewpoint or recollection. Conversely, bound diaries are more difficult to use as a tool on the jobsite with regard to distribution.

As discussed in Chapter 6, it is a good practice to require on-site subcontractors also to complete daily reports and provide a copy to the project manager and superintendent. In this way, the general contractor has in the subcontractor's original, written-in-hand evidence of their manpower and their view of daily progress. If there is a problem or restraint noted on a subcontractor's diary, the general contractor's project team should deal with it immediately. Some owners also may require copies of the subcontractors' daily reports as well as the general contractor's diaries.

## Control logs

Some of the types of logs that are used on a construction project include:

- short-form purchase order,
- requests for information,
- submittal,

- contract change directive,
- sketch,
- drawing,
- change order proposal,
- transmittal,
- letter,
- telephone conversation,
- subcontract modification,
- non-conformance report or quality control, and
- close-out.

Examples of many of these logs are shown in other chapters. To many beginning project engineers, the development and maintenance of the jobsite logs will be a major focus. All logs are important tools of the project management process. They provide summary reports of much of the communication processes on the jobsite and with the owner, design team, subcontractors, and suppliers. Although the production of the logs may be a project engineer's responsibility, it is still the ultimate responsibility of the project manager to ensure that they are correct, similar to files and meeting notes. Many of the logs are used for reference, others as a summary but, most important, logs are used as the focus of discussion during coordination meetings. Many of the logs reviewed during a meeting should then be attached to the meeting agenda and subsequent meeting notes. As will be discussed in Chapter 10, some of these logs may be maintained on the project website to provide visibility to all of the team members.

## Record drawings

There can be several different types of record drawings. The set of drawings estimated from and used and referenced in the original contract is kept as a set of record drawings. The current set of contract drawings, as incorporated into the contract through change orders, is a record set of drawings. The permit set of drawings as stamped by the municipality is a record set of drawings and is required to be available in the contractor's office for review by the inspector. Approved submittal drawings are to be maintained as record drawings.

An example of a portion of a drawing log for the NanoEngineering Building is shown in Box 9.2. This log may be maintained by the design team or the construction team or both. It should be periodically distributed to make sure that all contracting parties have the current set of drawings and that these drawings have been incorporated into all contracts, subcontracts, and purchase orders. The requirements for posting and storage of different types of record drawings are usually defined in the specifications or in the special conditions to the contract. The record drawing requirements for the NanoEngineering Building are contained in Section 00-72-00.4.02 of the specifications; an excerpt follows:

> Contractor shall legibly mark in ink on a separate set of the Drawings and Specifications all actual construction, including depths of foundations, horizontal and vertical locations of internal and underground utilities and appurtenances referenced...

# Box 9.2    Drawing log

### DRAWING LOG
**Date:** September 14, 2015
**Sheet:** 1 of 5

Project:    NanoEngineering Building                Job Number: 9821

| Document Number | Description | Rev 0 | Rev 1 | Rev 2 |
|---|---|---|---|---|
| A0.00 | Cover Sheet | 8/21/15 | | |
| A0.01-2 | Master Sheet Index | 11/10/14 | 8/21/15 | |
| A0.02 | Vicinity Map | 11/10/14 | 8/21/15 | |
| A1.01 | Site Plan | 8/21/15 | | |
| LS.01 | Life Safety, Lower Levels | 10/24/14 | 11/10/14 | 8/21/15 |
| LS0.2 | Life Safety, Upper Levels | 10/24/14 | 11/10/14 | 8/21/15 |
| A2.01 | Floor Plan Level B1 | 11/10/14 | 8/21/15 | |
| A2.03 | Floor Plan Level 1 | 11/10/14 | 8/21/15 | |
| A2.04 | Floor Plan Level 2 | 11/10/14 | 8/21/15 | |
| A3.10 | Building Section, E-W | 11/10/14 | 8/21/15 | |
| A3.11 | Building Section, N-S | 11/10/14 | 8/21/15 | |
| A4.01 | Exterior Details | 11/10/14 | 8/21/15 | |
| A7.01 | Interior Elevations | 8/21/15 | | |
| LF1.01 | Lab Furnishings, General Notes | 8/21/15 | | |
| LF1.02 | Lab Furnishings, Schedules | 8/21/15 | | |
| S0.01 | Structural Notes | 10/24/14 | 11/10/14 | 8/21/15 |
| S2.01-2 | Foundation Plan, Level B | 10/24/14 | 11/10/14 | 8/21/15 |
| S2.04-2 | Framing Plan Level 2 | 10/24/14 | 11/10/14 | 8/21/15 |
| M0.01 | Mechanical Symbols | 10/24/14 | 11/10/14 | 8/21/15 |

Another type of record drawing is the set of as-built drawings, which are also discussed in Chapter 17. As-built drawings are developed by marking up exact construction conditions on a copy of the construction drawings. These will be used by the owner as a reference of the manner in which the project was actually constructed. The superintendent is responsible for maintaining the as-built drawings to ensure they reflect any changes made during the construction of the project, but usually with the help from a PE or if all of the documents are on CAD or building information model (BIM), with a computer technician.

## Oral communications

Oral communications are important and are often essential. Typical oral communications include making presentations and informally discussing issues with other project team members, either individually or collectively in meetings. Written documentation should also be used to document oral communications. The telephone memorandum, illustrated in Box 9.3, is a standard tool extensively used in the construction industry. A copy should always be routed to those on the other end of the conversation as well as to any others involved in the subject. This is sometimes viewed negatively, but if the practice is started early in the job, it will soon be accepted by all as sound business practice. A copy of each telephone memorandum should be routed to the relevant team members and the original placed in the project file. Meeting notes also record oral communications as will be discussed in Chapter 10.

## Electronic communications

Electronic mail (e-mail) has completely changed construction communications. E-mail is an excellent, but somewhat impersonal, substitute for telephone calls and provides a written record of each communication, substituting for the telephone memorandum shown in Box 9.3. Electronic mail is being used to coordinate RFIs with designers and subcontractors, coordinate material deliveries with suppliers, transmit submittals, and coordinate project issues with owners. Electronic mail does not, however, eliminate the need for periodic coordination meetings, formal letters, and personal phone calls.

It was only about a decade ago that computer-aided design (CAD) completely revolutionized the design industry. Today, BIM has surpassed CAD with its capabilities for the architect to communicate three-dimensional concepts early to the owner and the city. BIM is also being used by many construction teams on the jobsites, especially on larger work and on those projects where the model was already originated by the design team. Contractors have added fourth and fifth dimensions to the model by inserting time and cost to the different design elements and phases. BIM is a valuable tool to coordinate design changes and resolve discrepancies before they significantly impact construction productivity. Some larger projects have a standard BIM coordination meeting outside of the regular weekly OAC meeting.

## Summary

Documents and document control are important tools for the construction team. The project manager is responsible to ensure that these systems are set up properly at the beginning of the

## Box 9.3   Telephone memorandum

### NORTHWEST CONSTRUCTION COMPANY
1242 First Avenue, Cascade, Washington  98202
(206) 239-1422

| TO/WITH: | DATE: *8/20/15* |
|---|---|
| *Robert Smith* | TIME: *11:15 am* |
| COMPANY: | JOB NO.: *9821* |
| *University of Washington* | PAGE: *1 of 1* |

**MEMORANDUM:** ☐ Conference   [x] Telephone ☐ File ☐ Information ☐ Fax

PHONE NO.: _206-929-4923_          FAX NO.: _____

PROJECT: _NanoEngineering Building_     SUBJECT: _Contract Exhibit_

LOCATION: _15ᵗʰ Avenue, Seattle, WA_

DISCUSSION:
_Robert called me this morning and asked when we would revise the Document
Exhibit included with the contract delivered in February and executed in March. I
indicated that although the design team is currently revising the drawings and
specifications to reflect City Permit comments, I did not anticiapte any major cost or
schedule impacts. Once these documents are revised, and we have had a week or so
to review, we will revise the Exhibit and then incorporate it and these new
documents into the contract via formal change order._

FILE CODE: 9821/OC                    BY: *Ted Jones*
COPIES: *Robert Smith/UW, Norm Riley/ZGF*

---

project. The project manager may delegate the maintenance of these systems to a PE or assistant, but it is still the project manager's responsibility to make sure that the systems are operating properly.

The files for a construction project should be developed consistent with other projects of the construction company. The files should be kept current and located where they can be readily accessed by the construction team. The daily job diary and the project files are valuable historical tools in case of a dispute. The diary should be maintained by the project superintendent. There are numerous logs that are used on a construction project. They provide a summary of the construction documentation and are often reviewed during construction meetings and attached to meeting notes as exhibits.

There are also various types of record drawings. The requirements for maintenance and storage generally are defined in the specifications. The contractor will want to store different sets of contract or reference drawings for extended periods after the project is completed for historical reference.

## Review questions

1. Which member of the construction team is responsible for physically mobilizing onto the jobsite?
2. Which member of the construction team is responsible for mobilizing the paper flow?
3. The file code system on a jobsite should be similar to what other code systems?
4. Where should the project files be located?
5. How long are files kept in storage?
6. What is a journal or daily report?
7. List at least five important types of information that is recorded on a daily job diary.
8. Who should originate a diary?
9. List three misuses of a diary.
10. List three different types of logs
11. Why would it be important for a document-control log to be up to date?
12. How would it be possible for different members of the team to be working off different revisions of construction drawings?
13. How are as-built drawings created?
14. List three types of record drawings.

## Exercises

1. On the NanoEngineering Building, what are the dates and the revision number for Drawing A2.03-2 that the drywall subcontractor should use as a reference in its daily diary when framing the first-floor elevator lobby?
2. Create a typical job diary for the NanoEngineering Building for the following dates. Use the project schedule on the website for reference. Make any reasonable assumptions:
   a.  January 21, 2016 (a snowy day);
   b.  August 21, 2016 (a clear day).

# 10 Communications

## Introduction

Communication broadly means acquiring and transmitting information. It is the most critical project management tool. A good technical project manager who knows how to estimate, plan, schedule, and execute may fail, unless he or she also has good communication skills. Unless the project manager can communicate his or her needs, wants, and expectations, they probably will be unfulfilled. Construction communications involve exchanging information regarding a project. Several formats and techniques have been developed to expedite the flow of information among members of the project team. Many communication tools such as contracts, schedules, logs, and start-up documents have been introduced in previous chapters. This chapter describes the use of several additional communication tools that are essential to the project manager.

## Web-based communication tools

Traditional paper tools, as discussed in this and other chapters, have in the past decade or so slowly become obsolete in favor of electronic communication tools. Computer and web-based project management systems such as ProLog, building information model, BlueBeam, Drop Boxes, and others have become the norm. However, these electronic processes have not replaced the concept of construction management tools such as requests for information (RFIs) and submittals and meeting notes but more how the document is transmitted, that is, electronically and not photocopied, faxed, or mailed.

## Traditional construction management communication tools

To be a skilled communicator, the project manager must clearly understand what message he or she is trying to send, how the message will be interpreted by the recipient, and what action will be initiated as a consequence of receiving the message. Communications must be clear and concise to avoid confusion. Simple language is more effective than cryptic jargon. Objectivity is preferred over subjective opinions and emotions. Acquiring good writing and speaking skills requires patience and practice but is required if one is to become an effective project manager.

Effective written communications are essential. Most people have short memories and often hear what they want to hear. People may hear, but interpret, the same communications differently. The project manager does not want to stifle verbal communications, teamwork, and camaraderie for the sake of the written word, but there is a need to make a written record of discussions affecting a project. Written forms are used to document verbal communications, such as meetings and telephone conversations. They form a clear record of what happened and what was intended or directed to happen. Unfortunately, some people see written forms of communications only as documentation to support claims or disputes. This does not have to be the case. If project communications are properly handled, there should be fewer disputes and claims. Written communications should be looked upon as construction management tools to help build the project, not solely as support for future litigation.

Transmittals are one of the most common construction communication tools. An example is shown in Box 10.1. This document works as the cover sheet when delivering other types of documents. For example, a transmittal may be used to send a request for payment to the owner. It may be used to send a revised schedule to a subcontractor. Transmittals need to clearly tell recipients what they are being given, what they are expected to do with the attached documents, and when they are to do it. Transmittals should be completed using simple, direct language to avoid confusion. The original is sent to the recipient with the documents being transmitted. A second copy is placed in the project file to serve as a record of what was sent and when. A third copy could be sent to either the project superintendent or the officer-in-charge, so that he or she also is apprised of what documents are changing hands. Or all correspondence may be electronic.

Some other common types of written construction communication documents include:

- contracts (Chapter 2),
- estimates (Chapter 3),
- schedules (Chapter 4),
- submittals (Chapter 7),
- memorandum and letters (Chapter 9),
- job diaries (Chapter 9),
- electronic mail (e-mail),
- meeting notes,
- RFIs or field questions,
- project team lists (companies, individuals, telephones, and addresses),
- change orders (Chapter 15), and
- warranties (Chapter 17).

As shown, most of these documents are discussed in detail in other chapters. The major issue is that they be prepared in simple, direct language to eliminate the potential for misinterpretation.

# Box 10.1   Transmittal

## NORTHWEST CONSTRUCTION COMPANY
1242 First Avenue, Cascade, Washington  98202
(206) 239-1422

### TRANSMITTAL

Date: *February 11, 2015*

To:   *University of Washington*
*15th Avenue N.E.*
*Seattle, WA 98195*

Attn:  *Robert Smith, Owner's Representative*

Re:   *Contracts*

Enclosed: X      Fax: _____   Mailed: _____   Prior, w/o transmittal: _____
Hand Delivered: X  Fax Number: _____

Item(s) Transmitted: *2 each original ConcensusDocs® contracts #500, dated*
*February 11, 2015, with attachments*

Remarks:   *We have completed the contracts as agreed and have executed both*
*original copies.  Please initial each page, including attached exhibits,*
*and return one completely executed original for our files.  If you have*
*any questions regarding the enclosed, please call.*

Transmitted by:      *Ted Jones, Project Manager*

Job/File code:  *9821*

cc: *Mr. Sam Peters, NWCC*
*Mr. Norm Riley, ZGF Architects*

## Project meetings and meeting minutes

The primary purpose of a meeting is to provide a forum for direct communication and timely and efficient exchange of information. Creating action and obtaining decisions necessary to maintain the scheduled flow of work enhance that communication. The meeting itself is a tool that is to be used to assist in constructing a project.

## Types and frequency of meetings

Informal meetings are held every day. If two people pass in the hall, stop, and exchange information, a meeting has occurred. Our discussion will be focused on the more formal scheduled meetings. The project manager will be involved with various types of meetings, each with its purpose, attendees, format, and notes arranged according to the subject of that particular meeting. The project manager's role also varies with regard to the subject of the meeting. Some of the types of meetings that occur on a typical project are shown in Table 10.1. Owners may include meeting requirements in their contract

**Table 10.1**   Meeting types

| Title | Occurrence | Chair |
|---|---|---|
| Owner/Architect Meeting (OAC) | Weekly | Project Manager |
| Safety | Weekly | Assistant Superintendent or General Foreman |
| Scheduling Preconstruction meetings: | As Needed | Project Manager or Superintendent |
| •  with Owner and Architect | Once | Owner or Architect |
| •  with Subcontractors | Once per subcontractor | Project Manager |
| •  with Unions (Jurisdictions) | Once | Superintendent |
| •  with City (Permits, utilities) | Once | Project Manager or Superintendent |
| •  coordination meetings | As Needed | Project Manager or Architect |
| •  design coordination | Weekly | Architect |
| Labor Relations | As Needed | Superintendent |
| Foremen's meetings | Weekly | Superintendent or Assistant Superintendent |
| Change order status/review | Weekly | Project Manager or Architect |
| Pay request review | Monthly | Project Manager or Architect |
| Punchlist meeting (end of job) | Weekly | Project Manager, Superintendent, or Architect |
| Subcontractor coordination | Weekly | Project Manager |
| RFI/Submittal coordination | Weekly | Project Manager or Project Engineer |
| BIM coordination | Weekly | Project Engineer |

specifications or contract. Article 3.3.2 of the ConsensusDocs 500 contract on the companion website contains the preconstruction meeting requirements for the NanoEngineering Building.

Meetings should be announced in sufficient time so that all responsible parties can arrange to attend. A meeting notice, similar to the one shown in Box 10.2, works well for this purpose. Meetings should be held in the office of the firm conducting the meeting. The project manager should chair all meetings that are appropriate. Only those parties with pertinent information should attend. Too large a group causes simultaneous third-party conversations, although additional attendees are often necessary to hear the directions given first hand. A short electronically issued memorandum is sometimes necessary to remind the required attendees of their expected continued attendance. The meeting notes, discussed below, will provide the appropriate notice for the next meeting. Meeting attendance should be taken, with a list of attendees attached to the meeting notes. This reminds attendees of the importance of their presence and serves as a permanent record of their attendance.

## Meeting agenda

If a particular meeting is not a continuation of a previous meeting, an agenda is required. The agenda is a sequential listing of the topics to be addressed at the meeting. This document forces the meeting to proceed in a formal manner that allows the project manager the opportunity to accomplish his or her goals and limits digression. The project manager first establishes himself or herself as the meeting leader or chair by sending out the agenda and meeting notice. It is a good procedure to distribute this agenda at least one day prior to the meeting, allowing all parties to prepare properly. If this is not possible, a printed agenda distributed at the beginning of the meeting is still beneficial.

## Conducting the meeting

The weekly owner/architect/contractor (OAC) meeting, also known as the *project coordination meeting*, is the project manager's primary means of communication with the entire project team at one time and in one place. It is held either in the general contractor's main office or in the field office depending upon project progress and size. This meeting ideally is held at the jobsite trailer, which helps instill the importance of timely decisions upon the team members. It is imperative that both the architect and the owner attend this meeting.

The meeting chair should make every endeavor to follow an agenda. All items requiring discussion and resolution should be addressed. If a meeting wanders from the agenda or contains multiple simultaneous conversations, it will last excessively long, causing issues to be tabled, postponing action items, and ultimately may even delay the project. It is preferable to hold off individual conversations until after the meeting has ended. Workshops, break-out sessions, or task forces can be utilized effectively for this purpose. The smaller group can then report back to the larger group at the next coordination meeting. The project manager must avoid being too rigid by not allowing discussion of new items, which may also be important. If attendees are not getting what they want out of the meeting, their attention and attendance may diminish. The project manager is also the peacekeeper, head communicator, and lead compromiser and must lead by example in all cases. He or she must have the ability to bring discussions to a close and identify specific actions to be taken regarding open issues.

## Box 10.2    Meeting notice

### NORTHWEST CONSTRUCTION COMPANY
*1242 First Avenue, Cascade, Washington  98202*
*(206) 239-1422*

**MEETING NOTICE**

Date: *March 20, 2015*

To:    Required Attendees
      Optional Attendees

Subject: *The city has asked to meet with the owner and construction teams to discuss traffic flow on 15th Avenue in light of several new construction permit applications in this area.*

Date of meeting:                  *March 30, 2015*
Meeting start time:           *9:00 a.m.*
Anticipated completion:     *10:30 a.m.*

Location of meeting:       *General contractor's project site trailer*

Required attendees:                      Representing:

*Ted Jones, Jim Bates, Mary Peterson*    *Northwest Construction Company*
*Norm Riley*                           *ZGF Architects*
*Robert Smith*                       *University of Washington*
*Mary Jackson*                     *City of Seattle Permits*
*Jerry Hopkins*                    *City of Seattle Traffic Engineer*

Optional attendees:                      Representing:

*Frank Waters*                     *Civil Engineering Consultant*
*Neighbors*                          *Notice posted and distributed in neighborhood*

Attendance comfirmation required?    Yes: _X_  Please call: (206) 239-1422.
No: __

Transmitted by:    *Mary Peterson, Project Engineer*

Job/File code: *9821/MN*

Choice of seating is important. The important decision makers should not be standing against the wall or sitting in the second row. There are two seats of choice for the project manager:

- **The head** of the table establishes a position of power and leadership. If the project manager has a message to deliver, this may be the best chair. If the project manager has not established himself or herself as the leader of the group, this also may be the best chair.
- **The middle** of a long side of the table offers the project manager the best opportunity to communicate with many people. If the goal of the meeting is to achieve a compromise, the project manager may choose this softer or friendlier position.

As indicated previously, the coordination meeting should follow an agenda. Use of the previous meeting's notes provides an easy and natural path to follow. The following general outline works in most situations:

- status of the project schedule (the superintendent's responsibility);
- safety issues (the superintendent's responsibility);
- review of various document logs (often presented by the project engineer);
- review open items or "old business" from the previous meeting;
- new business items for discussion: all participants given an opportunity to introduce new items, starting with the GC's PM;
- summary of decisions made; and
- announcement of the next meeting date and time.

Some project managers summarize many of the important issues decided upon during the meeting or actions due prior to the next meeting. This is done verbally to all in attendance at the end of each meeting. This process is similar to reading the first draft of the meeting notes. This is a good reminder and offers all the opportunity to disagree if their understanding is not the same as the project manager's.

# Meeting notes

The uses of meeting notes are multi-faceted. They include:

- providing a formal list of meeting attendees;
- providing a written record of meeting discussions and decisions;
- creating responsibility through action items with due dates;
- reminding attendees of action required prior to the next meeting;
- providing an agenda for next meeting;
- soliciting corrections to the project manager's understanding of meeting results; and
- announcing next meeting.

It is the project manager's responsibility to chair the weekly project coordination meeting and publish meeting notes. A meeting may be too large for the project manager to both chair and record meeting notes. In such instances, a PE or assistant project manager may assist with taking the notes. The project manager should always take and publish his or her own set of notes, even when the meeting has been called by, chaired by, or hosted by another party. The note publisher will, if only subconsciously, tend to slant the notes toward his or her desired outcome. Notes should be prepared and distributed on the same day the meeting is conducted. This minimizes the possibility of memory loss and provides a timely notification to attendees of their commitments and actions due. The project manager may telephone or e-mail attendees a couple of days later and verify that they received their notes, question them regarding accuracy, status open-action items, and remind them of the next meeting. Project managers should not take the position that just because the notes were mailed, or most likely e-mailed, that they were received and read.

Meeting notes serve as a formal and sometimes legal record. They are important documents that should be regarded seriously. Meeting notes often serve as supportive data for letters, change orders, and RFIs. They should be filed and stored as part of the project files. The format for meeting notes should follow the basic format shown in Box 10.3. Meeting notes may occasionally vary according to specific projects, topics, and attendees. Individual project meeting notes used within a construction firm should follow the same format throughout the firm. At a minimum, the notes should contain the following:

- heading: project name, construction firm's letterhead, date and time of meeting, meeting number, location of meeting;
- typed attendance list and the firm each attendee represents;
- paragraph notifying attendees that they are expected to be qualified to act for the firms they represent and that the decisions they make represent that of their firms;
- paragraph notifying participants and note recipients that they have up to 7 days to notify the note taker of any errors or omissions. after that time, the notes will be considered statements of fact and entered into the project records;
- item numbers: each new business note item shall be numbered with that meeting number as its prefix. this will eliminate duplicate numbers and allows next week's notes to tie in. also open or unresolved items can be carried forward as old business. the meeting number prefix also indicates the age of open issues;
- discussion, action items, and decisions;
- company or individual responsible for taking action;
- dates the action is to be due and the individual responsible for making it happen;
- full sentences used wherever possible;
- the use of **bold**, or *italics*, or <u>underline</u> are beneficial to bring attention to topics;
- reference, and if at all possible, attach any handouts and logs, regardless of source; this way the project manager can avoid one party indicating it did not receive a copy of an item distributed during the meeting;
- note taker's name;
- distribution to all attendees, those absent, and others whom the notes affect;
- attached logs distributed and discussed, such as RFI log, submittal log, ASI or sketch log, and change order proposal log;
- each page ; numbered, for example, page 2 of 4; and
- underneath the last discussion held in the meeting: "end of meeting"; and
- every document on a construction job filed and assigned a file code number that includes the job number; this code goes on the bottom of the last page of the meeting notes.

# Box 10.3   Meeting notes

## *NORTHWEST CONSTRUCTION COMPANY*
1242 First Avenue, Cascade, Washington  98202
(206) 239-1422

**OAC Construction Coordination Meeting Notes**
Meeting Number:   20
Date:  September 14, 2015
Sheet: 1 of 5

| Project: | NanoEngineering Building | Job Number:  9821 |
|---|---|---|

| Attendance: | Representing: |
|---|---|
| Robert Smith | University of Washington |
| Norm Riley | ZGF Architecture |
| Ted Jones | NWCC |
| Jim Bates | NWCC |
| Mary Peterson | NWCC |

Representatives attending this meeting shall be qualified and authorized to act on behalf of the entity that each represents. To the best of our knowledge, this is an accurate summary of the discussions, decisions, etc. which occurred during this meeting. Notification of exceptions to this summary are to be made within seven days of receipt. Please bring your copy of these notes to the next meeting as they shall serve as the preliminary agenda.

Item    Discussion                                                                 Responsible  Date Due

Permits:
1.01    The elevator contractor is still responsible for obtaining their own permit
                                                                                ACME, 10/1/15

Schedule:
6.01    Jim reported that the project is within one or two days from the detailed contract schedule which is posted on the trailer's conference room wall. He impressed upon all the importance of turning around RFIs and submittals in an expeditious  manner so as not to impact the schedule.                                                   All      Ongoing

20.01  Jim handed out this week's short interval schedule. It is attached to these meeting notes. Norm requested copies of the subcontractors' three week schedules as well. Jim said that NWCC is receiving these as they are a subcontract requirement of the major subcontractors, but we will not be distributing them. Norm would be welcome to look through our schedule file if he had a specific question.

Safety:

## Requests for information

RFIs are asked for when a subcontractor or the general contractor discovers a discrepancy in the construction documents or with existing site conditions and needs assistance, or confirmation, from the designer or owner for resolution. They may be transmitted on paper or electronically. The RFI is a specific type of written communication that is used to document the issue and record the response. There are many terms used and associated abbreviations and acronyms for RFIs in construction. The most common are:

- design change verification request (DCVR);
- field information memorandum (FIM);
- field question (FQ);
- request for clarification (RFC);
- request for interpretation (RFI); and
- request for information (RFI).

While the format of each may be slightly different, the basic concept is the same. RFIs generally are managed by the project engineer (PE) who maintains a log indicating the status of each question. Questions generated by subcontractors or suppliers are submitted to the PE for review prior to their submission to the owner or designer.

## RFI preparation

RFIs can be originated by several of the construction participants. They may be written either by subcontractors, by suppliers, or by the general contractor. Regardless of who discovers the conflict or needs assistance, there are a few rules that should be followed when preparing an RFI.

- The question always should be asked in writing. Sometimes the PE contacts the structural engineer verbally with a critical question. These verbal questions should always be followed up requesting written confirmation of the discussion held.
- Subcontractors and suppliers should always route questions through the general contractor. Many subcontractors use their own forms. The project manager may allow subcontractors to prepare RFIs on their own forms and then use the general contractor's form as a cover with his or her own numerical identifier. Some project managers require subcontractors to use the general contractor's form. The exact form used is not as important as the information it should contain. Some contract specifications may dictate the form to be used on a specific project.
- The number of questions asked on a single form should be limited. If too many questions are simultaneously asked, the designer will wait to respond until he or she has answers to all questions. This will delay the response.
- Each question must be clear, complete, concise, and accurate. It must be referenced to specific contract sections and supported with appropriate documentation. Such documentation may include photographs, sketches, and material data.
- Each form should have a unique number for identification.

An example RFI for the NanoEngineering Building is shown in Box 10.4. This specific form was customized by our case study general contractor. Other generic RFI forms incude ConsensusDocs 204 and AIA G716. Other architects or clients may have their own customized forms they wish the contractors to use and will be included in the contract special conditions.

All RFIs should be tracked on a log. Box 10.5 shows an example. This log and others, such as submittal and sketch logs, are generally maintained by the project engineer. It is important to note who generated the question as well as who is expected to respond. Open items on the log should be reviewed during the OAC coordination meeting. This log should be an attachment to the meeting notes, as discussed previously.

Verbal responses to RFIs should not be accepted. If the reviewer refuses to write a response, the project manager should write the designer's verbal direction regarding the question in the answer portion, document the source and time, and route the form back through, requesting confirmation, similar to copying the other party with a telephone memorandum.

All responded RFIs should be routed to the appropriate office and field staff as well as the originating subcontractor. They also should be sent to any other subcontractors and suppliers who may possibly be involved or impacted. All responded questions need to be reviewed for project impact; they may change project scope or impact the schedule. RFI impact analysis is discussed later in this section.

## RFI management

Like submittals that were discussed in Chapter 7, RFIs are part of the final stages of the design process. Many conflicts could not have been anticipated until the duct and pipe are actually installed on the jobsite. Many unforeseen conditions are exposed when the backhoe excavates for the foundations. If the RFI process is used properly, it can be a valuable tool for the construction team. Many architects and owners dislike the RFI process. It costs time and money to respond. They feel many RFIs are superfluous or the contractor is just digging for change orders. Unfortunately, some contractors use RFIs for just that purpose. If the issue is identified or anticipated early enough, the designers do not feel like they are always being asked for an immediate response. Many designers who realize that the construction and design processes are not independent disciplines also acknowledge that well-written RFIs from the contractors, installers, craftspeople, and fabricators are really an active process of quality control assistance rather than annoyance.

Some construction participants advocate conducting a meeting to discuss several questions at the same time. This often is unavoidable and can be helpful. Sometimes the only way to visualize the conflict is for everyone to meet on the project site. Meetings such as these also can have a positive impact on team relationships. The resolutions should still be documented, either during or after the meeting.

Designer originated question forms often have spaces for the contractor to indicate if the issue will result in cost or schedule impacts to the project. This is a difficult situation for the project manager. The issue, the conflict, or the question, should be resolved first, best, and expediently, rather than force the designer into a position in which he or she must taint the answer because of a potential change order and possible retaliation from the owner. If the issue is always cost first and constructability later, the RFI's use as a tool is diminished.

## Box 10.4    Request for information

### NORTHWEST CONSTRUCTION COMPANY
1242 First Avenue, Cascade, Washington  98202
(206) 239-1422

### REQUEST FOR INFORMATION

Project:        *NanoEngineering Building*          Date:  *May 20, 2016*
Area/System: *Elevator Lobby*                      RFI No. *241*
To:             *ZGF Architects*                   Related RFIs: *NA*
Address:        *925 Fourth Avenue, #2400*
                *Seattle, WA 98104*
Attention:      *Norm Riley*
Required Response Date:   *May 27, 2016*

Subcontractor/Supplier Forwarding Question: <u>*NA*</u>     Sub's RFI No.  <u>*NA*</u>

Subject:        *Exposed Elevator Lobby Ceiling*

Detailed Description/Request:

*The Fire Marshal visited the site at our request to make sure our construction operations were meeting his expectation, including temporary fire extinguisher locations. We also discussed when he was expecting the fire sprinkler system to go live. During this walkthrough, he noted the exposed elevator lobby ceiling on the first floor. He indicated that this area was directly connected to the egress hall and that he was going to expect that a one-hour hard lid be installed here. We did not provide comment or code interpretation but indicated we would forward his comment to the Project Architect for review. He did not leave any inspection report. Please review and advise.*

Attached/Referenced Drawings/Photographs/Specifications:    *A2.03-TI*

Cost or Schedule Impacts:  *Unknown at this time*

Please reply to:    *Mary Peterson, Project Engineer*

Architect/Engineer Response:      <u>*The Fire Marshal is correct. We will develop a revised finish schedule and forward before the end of this week. See CD #7*</u>

Signed:    *Norm Riley*                      Date: <u>*May 22, 2016*</u>

File Code:    *9821/RFI*

# Box 10.5    RFI log

## RFI LOG

Project Number: _9821_     Project Name: _Nano Engineereing Building_     Project Manager: _Ted Jones_

| RFI | Originator | Sub RFI | Support Docs | Description | Submitted | Requested | Returned | COP |
|---|---|---|---|---|---|---|---|---|
| 239 | Client | | Phone Memo | Paint Changes | 4/30 | 5/15 | | |
| 240 | NWCC | | Sketch, Photo | Exterior Walk Dimension | 5/1 | 5/7 | 5/7 | NA |
| 241 | NWCC | | A2.03-TI | Elevator Lobby Ceiling | 5/20 | 5/27 | 5/22 | 55 |
| 242 | Richardson | 15 | | Change PVC for EMT in Garage | 5/20 | 5/27 | | |
| 243 | Richardson | 16 | Cut Sheet | Verify wattage of lamp specifed | 5/31 | 6/7 | 5/31 | NA |
| 244 | Arctic | 37 | BIM SK 77 | Beam and Duct conflict | 5/31 | 6/1 | | |

RFIs are part of the active quality control process. If there is a problem, it should be addressed, not covered over. Project managers cannot afford to take the chance that they may think they know how the owner or designer will answer the RFI, and, therefore, they do not need to ask it. How many RFIs can a job have? It depends upon the parties involved, the type of contract, the complexity of the job, and the value of the contract. It also depends upon the quality of the design documents. The better and more complete and coordinated the design, the fewer RFIs that will arise in the field. Project teams that have worked together previously tend to have fewer questions. Negotiated projects also tend to have fewer RFIs than do competitively bid lump sum projects. It is important to follow correct project management procedures to ensure a successful outcome, especially on projects that require many field clarifications.

The PE (and subcontractors) must be as clear as possible when asking RFIs. The specific problem must be clearly stated in simple language. Photocopied portions of drawings or specifications should be attached if needed to clarify the issue. Photographs of exact field conditions also often are attached. Test reports, letters, submittals, or other documentation also may be needed. Even if the reviewing party may already have some of this information, the question must be drafted so that it is completely self-explanatory. The drafter should do all that is possible to assist the designer in answering the question. All pertinent information must be included. If the PE knows of a good and logical solution, he or she should recommend it. However, the contractor cannot assume the architect's or engineer's design responsibilities.

Most important, the PE should research the problem and always be right. Designers are busy. At this stage of the project, they have often moved on to another design. Their construction administration budget may be nearly expended. The construction team should not ask irrelevant or superfluous questions and thereby waste the designer's time and money, challenge their patience, and jeopardize the team relationship that has been built. The project manager needs to ensure that the PE and the subcontractors have attempted to resolve the issue internally before submitting it to the designer.

All sketches issued by the design team to accompany an RFI response should be numbered, dated, and refer to a construction drawing. Most designers are using electronic sketches or 8.5- by 11-inch paper versions for minor drawing revisions. It is easier and quicker to produce sketches on this size paper rather than reissuing a new drawing. Sometimes the change is complex enough that a new drawing is necessary. Any sketches produced by different members of the design team should all be routed through the architect for sequential numbering. This is the same as subcontractors routing their field questions and submittals through the project manager. If the designer does not produce a log of all sketches, the project manager should. This log should be periodically distributed at both subcontractor and owner-architect meetings to test its completeness.

Also, along with RFIs and submittals, these sketches are of great value in beginning the as-built drawing process discussed in Chapter 17. They can be taped to the back of the previous drawing with a red circle annotating the impacted area and a reference made to the sketch number. For example, if a new wall layout is necessary in research laboratory 200, which is shown on drawing A2.04-TI, the sketch showing the change, say sketch number SK-49, is attached to the back of drawing A2.03-TI, the previous drawing. Drawing A2.04-TI then has a red circle around laboratory 200 with the note, "See sketch # 49 on previous page." Today most of these revisions are being handled electronically upon issue, by either the design or the construction team.

Revised drawings should also be noted on a drawing log as was shown in Chapter 9. The contract document exhibit should be revised, reissued, and re-incorporated into all contracts. This can be quite an administrative task, but if an impacted party does not receive a revised room layout, the quality control program has just failed. All sketches and revised drawings should be issued through the project architect and should be accompanied by a document (also numbered) that provides clear and concise direction as to what the construction team is expected to do. This direction may come in the form of one of the following documents:

- Architect's Supplmental Instruction (ASI), AIA Form G710
- Construction Change Authorization (CCA)
- Construction Change Directive (CCD), AIA Form G714
- Design Change Notice (DCN)
- Field Change Notice (FCN)
- Field Order (FO)
- Interim Directed Change (IDC), ConcensusDocs® Form 203
- For our case study: Change Directive

## RFI impact analysis

The response to each RFI should be reviewed carefully by the project manager to determine whether it results in any of the following:

- change in the scope of the project,
- additional construction costs, or
- adverse impact to the construction schedule.

If so, the project manager should request a contract change order from the owner. The document often used for this purpose is known as a *change order proposal* (COP). In some instances, the designer will recognize that the response will result in a change order, and the owner will issue a construction change directive (CCD). The use of COPs and CCDs will be discussed in Chapter 15. Both document types should be numbered sequentially, logged, and discussed at the OAC and subcontractor meetings. All involved should be informed. The CCD should provide clear direction to the construction team delineating required actions. Common direction choices include:

- proceed with the work; estimate to be provided later or "to-be-negotiated";
- proceed with preparation of an estimate; do not begin work until the estimate is approved;
- proceed with the work on a time and material basis; track actual costs; and
- architect to provide lump sum cost estimate.

All sketches, revised drawings, and CCDs are construction management tools, and many will become contract documents. They should be evaluated by the construction team and incorporated into the prime contract and all appropriate subcontract agreements by formal change orders.

## Summary

Construction communications involve exchanging information regarding a project. They may be written, oral, or electronic. Good communications are concise and focused to avoid misinterpretation. Transmittals are used to forward other construction management documents and provide a written record. Other common written documents include daily job diaries, RFIs, change orders, and meeting notes. Meeting notes are used to record the issues discussed and decisions reached at project meetings. They also list open action items and individuals tasked with resolving them. Meetings, especially OAC coordination meetings, are important communication opportunities for the project manager. Each meeting needs an agenda to provide focus and a note taker to prepare the meeting notes. The exact format for the meeting notes is not as important as the fact that the notes were prepared and distributed to all attendees. RFIs are used on the jobsite when the contractor is not sure of a detail or a dimension and requires supplemental interpretation from the design team. Depending upon the project size and the procurement method, there may be thousands of RFIs on a project that warrant attention from all the contracting parties.

## Review questions

1.  List five types of written communication tools.
2.  What is the most common form of written communication document?
3.  Who should chair the weekly jobsite safety meeting?
4.  Who on the project management team is often responsible for preparation of meeting notes?
5.  Who on the project management team is responsible for the accuracy of the OAC coordination meeting notes?
6.  List five uses of meeting notes.
7.  Where should the project manager sit at a meeting?
8.  What documents should be attached to the meeting notes and why?
9.  When should the meeting notes be distributed and why?
10. Why are RFIs used on construction projects?
11. Why do some designers resist RFIs?
12. Why do some designers welcome RFIs?
13. When is the RFI log reviewed?
14. Who should prepare an RFI?
15. Are verbal RFI responses sometimes acceptable? If so, how should they be ultimately documented?
16. Do all RFIs result in change orders?
17. What advantages do web-based project management systems offer over computer-based project management systems?

## Exercises

1.  Prepare a transmittal for Northwest Construction Company to submit its monthly request for payment to the owner of the NanoEngineering Building. Refer to Chapter 11.

2. Prepare a telephone memorandum of a conversation between the project manager and the electrical subcontractor on the NanoEngineering Building confirming the scheduled delivery of electrical switch gear. Refer to schedules located in other parts of this text and on the companion website.

3. Prepare a set of meeting notes for your classroom discussion as if it were a construction meeting.

4. As the roofing subcontractor on the NanoEngineering Building, prepare an RFI asking for direction regarding the curbs and flashing around mechanical penetratons. Refer to appropriate drawings and specifications. Include the time when you need a response.

5. Assume you found a conflict in the drawings for the NanoEngineering Building. The finish schedule on Drawing A10.01-TI shows the walls of conference room 291 to be off-white, but the interior elevation shown in Drawing A7.05-TI shows the walls to be light green, which will be more expensive and cause difficulty in providing a quality finish. Prepare an RFI to resolve the discrepancy.

# 11 Progress payments

## Introduction

One of the most important project management functions is to get paid for the work performed. A project manager may have all the tools necessary to earn a profit on a job, but if the owner does not pay for the work, the contractor will not be able to realize a profit. Some project managers do not acknowledge the importance of preparing prompt payment requests. This is especially true with many subcontractors. If a payment request is not submitted on time, the project manager will not likely get paid on time. Cash management is essential, or the general contractor may find that he or she is unable to pay suppliers, craftsmen, or subcontractors. Good cash management skills, just like good communications skills, are essential if one is to be an effective project manager.

## Contract types

On most construction projects, requests for payment are submitted monthly. The formats and times are specified in the general conditions of the contract. Payment procedures used on the NanoEngineering Building are contained in Article 10 of the ConsensusDocs 500 contract on the companion website. Regardless of the type of contract, many of the procedures are similar. In this section, we will discuss some of the pay request differences that result from different types of contracts.

Cost-plus contracts (including those with a guaranteed maximum price (GMP)) have the following characteristics:

- Project managers request payment based on actual expenses; he or she must have already received the invoices from subcontractors and suppliers.
- The project manager generally is required to submit subcontractor invoices and general contractor payrolls to the owner with the payment request.
- It is almost impossible to over-bill.

- Fees and any lump sum items, such as general conditions expenses, are billed using a schedule of values (SOV) and percentage completion.
- The project manager may be subjected to periodic owner audits of actual costs incurred.

Payment on a lump sum contract is based on an SOV and percentage completion. Front loading and over-billing can occur on this type of contract, as will be discussed in the next section. General contractor records are rarely audited in a lump sum contract. A unit price contract allows for payment based upon quantities actually installed. If the contractor is to be paid $2,000 per ton for structural steel installed and has installed 50 tons, then he or she will be paid $100,000, less any agreed-upon retention. This process is quite objective and can be facilitated by an outside quantity measurement individual, firm, or team. Payment on a time-and-materials contract is based on actual labor hours times a contract labor rate plus reimbursement for materials based on supplier invoices.

# Schedule of values

The first step in developing the pay request is establishing an agreed-upon breakdown of the contract cost, or SOV. Often the contract will require that an SOV be submitted for approval within a certain time after executing the contract, for example 2 weeks. This SOV should be established and agreed upon early in the job, well before the first significant request for payment is submitted but after all subcontracts have been awarded.

Development of the SOV starts with the summary estimate that has been corrected for actual buyout values. This is shown as the GMP cost column in the center of worksheet Worksheet 11.1. This would be the SOV used on a cost-plus contract because the general conditions and fee are listed separately. On a lump sum contract, the general conditions and fee would be distributed proportionately across all payment items as shown on the right side of Worksheet 11.1. The SOV that the project manager would submit for a lump sum contract is the far right column of the worksheet, which is titled "adjusted totals'; none of the other columns would be shown.

Some contractors try to combine costs into single line items. In this way, they believe they can possibly over-bill, or they can hide the true subcontract values from the owner. The SOV should be as detailed as is reasonable. The project manager should do all that is possible to assist the owner in paying completely and promptly. Nothing should be hidden. At a minimum, the former 16 CSI divisions, or relevant divisions from the new system, should be used as line items. Major subcontractors should be listed where possible. Separate building components, building wings or distinct site areas, separate buildings, phases, or systems should be listed individually in a detailed schedule of values. A narrow view or summarized SOV for a closed-book project might look like Worksheet 11.2. An abbreviated SOV might make it difficult for the project manager to sell the monthly pay estimate to the owner and the bank and is therefore in direct conflict with his or her ultimate goal, which is to be paid timely.

Even if the contract does not require early submission of the SOV for approval, it is good practice to do so. The project manager does not want any future arguments with an owner or architect over a payment request. The SOV should be submitted for approval, just as a door hardware schedule would. Most owners will appreciate the openness, and this may help with facilitating prompt payment as well as establishing a high level of respect and trust.

# Worksheet 11.1   Schedule of values worksheet

NORTHWEST CONSTRUCTION COMPANY
**SCHEDULE OF VALUES WORKSHEET**
NanoEngineering Building

| CSI Division | Description | GMP Value | % of Subtotal | GC & Fee Prorated | Adjusted Totals |
|---|---|---|---|---|---|
| 2 | Demolition | 850,000 | 1.6% | 141,573 | **991,573** |
| | Reinforcement | 850,000 | 1.6% | 141,573 | 991,573 |
| | Foundations | 1,000,000 | 1.9% | 166,556 | 1,166,556 |
| | Walls & Slabs | 6,205,000 | 11.9% | 1,033,481 | 7,238,481 |
| 3 | Concrete Subtotal: | 8,055,000 | 15.4% | 1,341,611 | **9,396,611** |
| | CMU | 450,100 | 0.9% | 74,967 | 525,067 |
| | Stone Veneer | 1,075,000 | 2.1% | 179,048 | 1,254,048 |
| 4 | Masonry Subtotal: | 1,525,100 | 2.9% | 254,015 | **1,779,115** |
| 5 | Structural & Misc. Metals | 803,030 | 1.5% | 133,750 | **936,780** |
| | Rough Carpentry | 457,500 | 0.9% | 76,199 | 533,699 |
| | Finish Carpentry | 480,000 | 0.9% | 79,947 | 559,947 |
| 6 | Carpentry Subtotal: | 937,500 | 1.8% | 156,146 | **1,093,646** |
| | Insulation | 835,000 | 1.6% | 139,074 | 974,074 |
| | Roof & Accessories | 1,095,000 | 2.1% | 182,379 | 1,277,379 |
| | Waterproofing | 575,000 | 1.1% | 95,770 | 670,770 |
| | Sheetmetal | 575,980 | 1.1% | 95,933 | 671,913 |
| 7 | Thermal & Moisture Protection Subtotal: | 3,080,980 | 5.9% | 513,156 | **3,594,136** |
| | Doors | 752,000 | 1.4% | 125,250 | 877,250 |
| | Glazing | 1,522,000 | 2.9% | 253,499 | 1,775,499 |
| | Door Hardware | 399,000 | 0.8% | 66,456 | 465,456 |
| 8 | Division 8 Subtotal: | 2,673,000 | 5.1% | 445,205 | **3,118,205** |
| | Drywall | 1,778,800 | 3.4% | 296,270 | 2,075,070 |
| | Painting | 590,000 | 1.1% | 98,268 | 688,268 |
| | Acoustical Ceilings | 722,000 | 1.4% | 120,254 | 842,254 |
| | Carpet | 954,000 | 1.8% | 158,895 | 1,112,895 |
| | Tile & Vinyl | 750,550 | 1.4% | 125,009 | 875,559 |
| 9 | Finishes Subtotal: | 4,795,350 | 9.2% | 798,695 | **5,594,045** |
| 10 | Specialties | 388,290 | 0.7% | 64,672 | **452,962** |
| 11–13 | Equipment | 2,740,350 | 5.2% | 456,422 | **3,196,772** |
| 14 | Conveying Systems (Elevator) | 842,000 | 1.6% | 140,240 | **982,240** |
| 21 | Fire Protection | 772,000 | 1.5% | 128,581 | 900,581 |
| 22 | Plumbing | 2,750,000 | 5.3% | 458,030 | 3,208,030 |
| 23 | HVAC & Controls | 11,800,923 | 22.6% | 1,965,517 | 13,766,440 |
| 21–23 | Mechanical Subtotal: | 15,322,923 | 29.3% | 2,552,128 | **17,875,051** |
| 26 | Line Voltage Systems | 5,000,000 | 9.6% | 832,781 | 5,832,781 |
| 27 | Communication & Security | 1,148,000 | 2.2% | 191,207 | 1,339,207 |
| 26,27 | Electrical Subtotal: | 6,148,000 | 11.8% | 1,023,988 | **7,171,988** |
| 31,33 | Earth & Shore & Utilities | 3,139,664 | 6.0% | 522,931 | 3,662,595 |
| 32 | Paving | 255,050 | 0.5% | 42,480 | 297,530 |
| 32 | Misc. Site & Landscaping | 755,000 | 1.4% | 125,750 | 880,750 |
| 31–33 | Sitework Subtotal: | 4,149,714 | 7.9% | 691,161 | **4,840,875** |
| | Subtotal w/o GCs and Fee: | 52,311,237 | 100.0% | 5,450,000 | 61,024,000 |
| | Jobsite General Conditions: | 5,450,000 | | | |
| | Fee & Insurance & Excise & Contingency | 3,262,763 | | | |
| | Subtotal GCs & Fee | 8,712,763 | | | |
| | **TOTAL GMP:** | 61,024,00 | | | |

# Worksheet 11.2   Summary schedule of values

**SUMMARY SCHEDULE OF VALUES**

Northwest Construction Company
242 First Avenue, Cascade, WA 98202
(206) 239-1422
**NanoEngineering Building**
4/1/2015

| CSI | Description | Total Costs |
|---|---|---|
| 2 | Demolition | $ 991,573 |
| 3 | Concrete | $ 9,396,611 |
| 4 | Masonry | $ 1,779,115 |
| 5 | Structural Steel | $ 936,780 |
| 6 | Carpentry | $ 1,093,646 |
| 7 | Thermal and Moisture | $ 3,594,136 |
| 8 | Doors and Glass | $ 3,118,205 |
| 9 | Finishes | $ 5,594,045 |
| 10 | Specialties | $ 452,962 |
| 11-13 | Equipment and Casework | $ 3,196,772 |
| 14 | Elevators | $ 982,240 |
| 21-23 | Mechanical Systems | $17,875,051 |
| 26, 27 | Electrical Systems | $ 7,171,988 |
| 31-33 | Sitework | $ 4,840,875 |
| **Total GMP:** | | **$61,024,000** |

The fee on cost-plus contracts can be invoiced as a percentage complete that matches the overall project completion. If the project is 60 percent complete, then 60 percent of the fee has been earned. Most owners will not take issue with this approach. General conditions on cost-plus contracts can be invoiced three different ways:

• actual costs incurred,
• percentage complete based upon work constructed, or
• straight line with equal payments for each month.

Approved change orders can either be spread across the SOV as they are received or be added to the bottom as separate line items. This second method generally is the easiest to administer, but this complicates the owner's ability to track subcontractor monthly and final lien releases. Reformatting the SOV each month is similar to re-drafting the construction schedule. The record of the original SOV record is lost, and it may create an issue with the owner or the architect.

Some construction companies advocate hiding the fee and general conditions, or front-loading them. This is more prevalent on a bid contract than with negotiated work. It is recommended that each line item, including the fee and general conditions, be listed just as they would in the project cost accounting system. The SOV should look like the contract estimate. Trying to explain during an audit or a claim situation why the cost of the foundations was stated as $1,000,000 in the pay estimate but was only $800,000 in the original bid will be difficult. Spreading, but still hiding, the fee and general conditions as a weighted average over the SOV may be fair, but it will be difficult for the owner to track lien releases, as will be discussed later in this chapter. This is true for both GMP and lump sum projects.

## Payment request process

The most common billing practice is to request payment at the end of each month. The project manager must gather all the costs and prepare the monthly request that is submitted to the architect or owner for approval. This process should start on the twentieth of the month. Subcontractors and major suppliers should be required to have their monthly invoices to the project manager at that time. Subcontractors often do not do a good job of managing their cash flow. The project manager needs to push them to submit their monthly billings. Some project managers have the attitude that if the subcontractors do not get their invoices in on time, that is the subcontractors' problem, and they will not get paid this month. While this may be contractually correct, it is counter-productive. The project manager must do all that is reasonable to ensure the subcontractors get enough cash to keep them from going broke, at least while they are on the project.

On the twentieth of the month, all the subcontractors should estimate what percentage of work they believe they will have complete and in place through the end of the month. The suppliers also estimate what they believe they will have delivered to the site through the end of the month. The project manager and superintendent also forecast what direct work they believe will be in place through the end of the month.

Subcontractors and suppliers will want to be paid for materials purchased for the jobs and delivered to their own warehouse but not yet on the jobsite. Maybe fabrication is necessary, as is the case with ductwork or structural steel, or maybe the supplier is ahead of schedule or the general contractor is behind. Payment for materials stored off the project site has its complications. Sometimes payment for such stored materials is unavoidable due to scheduling reasons. Occasionally, it may be financially beneficial for all parties. For example, the plumbing subcontractor was able to purchase the copper pipe at discount because it was purchased with materials needed for another larger project. In these cases, the project manager needs to be sure that his or her interests, and that of the owner, are protected. The material must be stored in an insured and bonded warehouse. This also requires a personal inspection and verification. Most general contractors, architects, owners, and lenders try to avoid paying for off-site stored materials.

The project manager collects all the forecasts, estimates, and requests and assembles a draft pay request to submit for approval. This involves estimating the percentage complete for each item on the SOV at the end of the month. These percentages are multiplied by the value of each line item to produce the SOV shown in Worksheet 11.3. The one shown is a customized form for the NanoEngineering Building. Many project managers use similar spreadsheets for their monthly SOV submissions. Alternatively AIA Form G703 or ConsensusDocs Form 239 may be specified in the contract as the required SOV worksheets.

## Worksheet 11.3    Pay request schedule of values

**NORTHWEST CONSTRUCTION COMPANY**
1242 First Avenue, Cascade, Washington 98202
(206) 239-1422

Billing Period: From: 04/01/16
To:    04/30/16

**NanoEngineering Building**
14

Request #

**PAY REQUEST SOV CONTINUATION SHEET**

Invoice #

1234

| A | B | C | D | E | G (D+E) | G.1 (G/C) | H (C-G) |
|---|---|---|---|---|---|---|---|
| Line Item | Description | Contract Value | Previous Complete | Complete This Mo. | Complete to Date | % Complete | To Go |
| 1.0 | General Conditions | 5,450,000 | 2,840,000 | 237,000 | 3,077,000 | 56% | 2,373,000 |
| 2.0 | Demolition | 850,000 | 850,000 | 0 | 850,000 | 100% | 0 |
| 3.1 | Reinforcement | 850,000 | 850,000 | 0 | 850,000 | 100% | 0 |
| 3.2 | Foundations | 1,000,000 | 1,000,000 | 0 | 1,000,000 | 100% | 0 |
| 3.3 | Walls & Slabs | 6,205,000 | 6,205,000 | 0 | 6,205,000 | 100% | 0 |
| 4.1 | CMU | 450,100 | 0 | 200,000 | 200,000 | 44% | 250,100 |
| 4.2 | Stone Veneer | 1,075,000 | 0 | 0 | 0 | 0% | 1,075,000 |
| 5.0 | Steel | 803,030 | 600,000 | 203,030 | 803,030 | 100% | 0 |
| 6.1 | Rough Carpentry | 457,500 | 457,500 | 0 | 457,500 | 100% | 0 |
| 6.2 | Finish Carpentry | 480,000 | 0 | 0 | 0 | 0% | 480,000 |
| 7.1 | Insulation | 835,000 | 0 | 0 | 0 | 0% | 835,000 |
| 7.2 | Roofing | 1,095,000 | 0 | 1,095,000 | 1,095,000 | 100% | 0 |
| 7.3 | Waterproofing | 575,000 | 575,000 | 0 | 575,000 | 100% | 0 |
| 7.4 | Sheetmetal | 575,980 | 0 | 200,000 | 200,000 | 35% | 375,980 |
| 8.1 | Doors & Frames | 752,000 | 0 | 0 | 0 | 0% | 752,000 |
| 8.2 | Glazing | 1,522,000 | 0 | 0 | 0 | 0% | 1,522,000 |
| 8.3 | Door Hardware | 399,000 | 0 | 0 | 0 | 0% | 399,000 |
| 9.1 | Drywall | 1,778,800 | 0 | 224,000 | 224,000 | 13% | 1,554,800 |
| 9.2 | Painting | 590,000 | 0 | 0 | 0 | 0% | 590,000 |
| 9.3 | Acoustical Ceilings | 722,000 | 0 | 0 | 0 | 0% | 722,000 |
| 9.4 | Carpet | 954,000 | 0 | 0 | 0 | 0% | 954,000 |
| 9.5 | Balance of floorcovering | 750,550 | 0 | 0 | 0 | 0% | 750,550 |
| 10.0 | Specialties | 388,290 | 0 | 0 | 0 | 0% | 388,290 |
| 11.0 | Equipment & Casework | 2,740,350 | 0 | 0 | 0 | 0% | 2,740,350 |
| 14.0 | Elevator | 842,000 | 200,000 | 0 | 200,000 | 24% | 642,000 |
| 21.0 | Fire Protection | 772,000 | 200,000 | 90,000 | 290,000 | 38% | 482,000 |
| 22.0 | Plumbing | 2,750,000 | 750,000 | 250,000 | 1,000,000 | 36% | 1,750,000 |
| 23.0 | HVAC & Controls | 11,800,923 | 1,700,000 | 1,166,000 | 2,866,000 | 24% | 8,934,923 |
| 26.0 | Electrical | 5,000,000 | 1,500,000 | 500,000 | 2,000,000 | 40% | 3,000,000 |
| 27.0 | Low-Voltage Electrical | 1,148,000 | 0 | 0 | 0 | 0% | 1,148,000 |
| 31.0 | Earthwork & Utilities | 3,139,664 | 2,894,664 | 0 | 2,894,664 | 92% | 245,000 |
| 32.1 | Paving | 255,050 | 0 | 0 | 0 | 0% | 255,050 |
| 32.2 | Misc. Site & Landscape | 755,000 | 0 | 0 | 0 | 0% | 755,000 |
| 90.0 | Fee, Insur., Excise. Cont. | 3,262,763 | 1,165,150 | 235,320 | 1,400,470 | 43% | 1,862,293 |
| | Original Total | 61,024,000 | 21,787,314 | 4,400,350 | 26,187,664 | 43% | 34,836,336 |
| 99.1 | Change Order No. 1 | 20,000 | 20,000 | 0 | 20,000 | 100% | 0 |
| 99.2 | Change Order No. 2 | 145,980 | 0 | 0 | 0 | 0% | 145,980 |
| | **Current Total** | **61,189,98** | **21,807,31** | **4,400,350** | **26,207,6** | **43%** | **34,982,31** |

About the twenty-fifth of the month, the project manager should hold a short informal meeting with the architect, the owner, and maybe the general contractor's superintendent and the bank at the jobsite to review this month's anticipated pay request. At this time, the pay request is presented as a draft for discussion and approval. If any of the approving parties has a problem with a particular line item or percentage, the project manager still has time to request an explanation from the submitting subcontractor or develop additional detail. A job walk during this meeting is extremely helpful to visualize the work completed or that will be in place by the end of the month. Only percentage completions are usually reviewed, not dollars. If necessary, subcontractor invoices can be attached to this draft for backup. This draft pay estimate and the meeting promote teamwork among the owner, the architect, and the general contractor's project team. Again, similar processes are recommended for both bid and negotiated projects to facilitate timely payment.

Some contracts may require that the monthly pay request review and approval coincide with review of the ongoing as-built drawing process. As will be discussed later, general contractors usually do not do a good job of developing as-built drawings. They often put it off to the end of the job and have them prepared by individuals who were not involved or who have forgotten what occurred. Monthly review of as-built drawings on the contractor's and subcontractor's jobsite plan tables, concurrent with the site walk and draft pay estimate, is an early active step in the close-out process.

The project manager submits the formal pay request, as revised (if necessary), for formal approval and payment to the owner no later than the end of the month. This can be submitted earlier if possible. Some contracts, and lenders, may also request that the architect sign off in the approval process. The project manager should carry the payment request through the approval process to ensure there are not issues with its content.

ConsensusDocs Form 291 and AIA Form G702 are popular copyrighted pay estimate summary coversheets; either of which may be stipulated as required in the specifications. Box 11.1 was developed by our case study client as another example of a pay request summary and approval page that accompanies the customized detailed SOV shown in Worksheet 11.3. Regardless whether one of these copyrighted documents or one created by the contractor or client is used, all blanks should be filled in. If any are not applicable, NA should be entered. The form is usually typed or may be prepared in ink.

The contract will dictate payment terms. It is customary for the general contractor to be paid by the tenth of the following month, if the payment request is submitted by the end of the current month. The specific timeframes used on the NanoEngineering Building are contained in Article 01-29-76 of the supplemental conditions. After the general contractor has received the monthly payment from the owner, he or she should disburse funds to the suppliers and subcontractors expeditiously. This usually is done within 1 week to 10 days after receipt. A project manager should not delay paying the suppliers and subcontractors.

Some third-tier subcontractors and suppliers may want to enter into joint check agreements with the general contractor to ensure that they are paid expeditiously. In this case, one joint check is cut simultaneously, naming both the primary subcontractor and its supplier or third-tier subcontractor. Although some project managers see this as a complicated task, it ultimately benefits and protects the general contractor and the owner from the potential that some subcontractors may not make their required payments and the third-tier firms ultimately lien the project.

# Box 11.1    Pay request summary

## PAY REQUEST SUMMARY
Where the basis of payment is a Guaranteed Maximum Price

Client:        *University of Washington*
Address:    *1101 15ᵗʰ Ave. NE, Seattle, WA 98011, (206) 239-3556*

Contractor:  *Northwest Construction Company*
Address:    *1242 First Ave, Cascade, WA 98202, (206) 239-1422*

| | | | |
|---|---|---|---|
| Project: | *NanoEngineering Building* | Project # | *9821* |
| Billing Period From: | *04/01/2016* | Request # | *14* |
| Billing Period To: | *04/30/2016* | Invoice # | *1234* |

| | |
|---|---|
| Original Contract Price: | *$61,024,000* |
| Approved Change Orders: | *$165,980* |
| Current Contract Price: | *$61,189,980* |
| | |
| Total work completed to-date, see attached SOV: | *$26,207,664* |
| Less retention held at rate of 5" | *-$1,310,383* |
| Plus applicable state sales tax at 9.95" | *$2,607,663* |
| Subtotal completed to-date, less retention, plus tax | *$27,504,943* |
| Less prior applications for payment: | *-$22,778,285* |
| Net amount due this pay period: | *$4,726,659* |

Contractor's Certification:
I hereby do certify that to the best of my knowledge the above accounts reflect a true and accurate estimate to the values of work complete and that all amounts paid prior have been distributed according to existing contractual requirements and the laws in existence at this jurisdiction at the time of the contract.
Contractor:  *Ted Jones*                          Date: *4/30/2016*

Architect's Certification:
In accordance with the contract documents and limited on-site evaluations, we have no reason to believe that the accounts presented by the contractor are not reflective of the work accomplished to date.
Architect:    *Norm Riley*                          Date: *5/1/2016*

# Cash as a tool

Cash is one of the most useful tools the project manager has. If a construction firm has a reputation of paying their subcontractors and suppliers quickly and fairly, they may receive more favorable prices on bid day. This will result in the contractor getting more work.

General contractors (the same applies here for subcontractors) are not banks. The general contractor's role on the construction project is to build the building, not provide construction financing. Each contractor begins incurring labor and material expenses on the first of the month. Bills for these expenses are submitted at the end of the month, and payment is not received until the tenth or the thirtieth of the next month. This means the general contractor has provided funding for these expenses for a minimum of 40 and maybe up to 60 days. These time periods may be up to 30 days longer for subcontractors. This is the reason why some project managers attempt to over-bill the owner.

If the project manager over-bills too much, the owner and bank may become suspicious, and trust and faith in the project manager are lost. The owner may allow a little over-billing so as not to place the contractor in a situation with too large of negative cash flow. The project manager should cooperate by not over-billing too much, as it might impede the monthly payment request process. The way some project managers over-bill is to front-load the SOV. Ways of accomplishing this are to place all the fee and general conditions (if not separately listed) on the early scheduled construction activities, such as foundations or concrete slabs. Another way is to artificially inflate these early activities or to falsify the amounts requested by subcontractors. These methods are also counter-productive to building with a team attitude and will most often ultimately be discovered through interim lien releases or during an audit. We do not advocate this process and have not included an example of a front-loaded SOV but have included an exercise later for the interested reader.

Another example of cash as a tool is the use of discounts. Discounts may be offered by material suppliers for early payment. If the general contractor pays within a certain time period, say 10 days, he or she can receive a discount off the invoice price. For this reason, a project manager may elect to pay suppliers early, even though the owner has not made a payment. In effect, the general contractor is acting as a bank. Many owners will want to benefit also from these discounts. If the owner has not paid the general contractor up front and the contractor is utilizing its own cash to realize the discount, the discount should belong to the contractor. If the owner is willing to pay up front and the general contractor is using the owner's cash to achieve the discount, then the discounts should apply to the owner.

## Retention

A portion of the monthly pay request is commonly held back by the owner (and subsequently by the general contractor with respect to subcontractors). This is referred to as *retention*, or *retainage*, and is another cash tool. The purpose of retention is to ensure that the contractor's attention is focused on completion of the project, including close-out of all paperwork activities. Discussion of close-out activities and the process required to release the retention is discussed in Chapter 17.

The amount of retention will be dictated by the contract and may be a subject for negotiation, or during the course of the project, re-negotiation. In the past, 10 percent was standard, but as the size of contracts has increased along with the value of money, 10 percent may be excessive in today's market. The most common percentages today are either 5 percent through the course of the job or, alternatively, 10 percent until the project is 50 percent complete and no further retention thereafter, which also equals 5 percent upon completion. Article 10.2.4 of the contract agreement for the NanoEngineering Building located on the companion website indicates a retention rate of 5 percent.

A subcontractor's contract is tied into the prime contract. In this way, the subcontractor is also tied into whatever retention terms the general contractor has with the owner. This is usually in the subcontractor's best interest and prohibits the general contractor from over-withholding.

One twist to retention is the length of time a particular subcontractor's money is held. If the owner retains 5 percent of the entire job until 30 days after substantial completion (the standard time frame), does this mean that the site work subcontractor has to wait to receive final payment until all the floor covering punch list items are complete? This is not fair to the early subcontractors and can result in undesirable financial burdens that can ultimately impact both the general contractor and the owner. It is a good idea to include text in the contract that allows for release of that portion of the overall retention to facilitate paying off these early subcontractors. The utility contractor can go broke on another project, and the project manager may receive a lien on this project because the retention is still outstanding. This is not as applicable on short-duration projects but is extremely important on jobs that last a year or longer. Early subcontractors should be paid and their final and unconditional lien releases received soon after their work is completed.

Sometimes on cost-plus contracts, the project manager can negotiate that retention is limited to a percentage held on the fee, for example 25 percent of the fee. Retention serves several purposes. It primarily establishes an account to finish the work of those parties who refuse or cannot do so, and it also serves as an incentive for the contractor to finish the project expeditiously. A project manager often may be able to convince the owner that he or she does not need to hold a large retention percentage, because the teams have worked together previously. Coincidently, 5 percent retention approximately equals a mid-size commercial general contractor's fee. Obtaining that retention is an incentive for the general contractor to have a smooth and expeditious close-out process as further discussed in Chapter 17.

## Lien management

A lien is a legal right to secure payment for work performed and materials provided in the improvement of land. This right attaches to the land in a manner similar to a mortgage. If a contractor is owed money and the owner refuses payment, the contractor can attach a lien to the owner's property. If the lien is not removed by payment, the lien claimant can demand legal foreclosure of the property and have the obligation satisfied out of the proceeds of the sale of the property. Liens also make it difficult for a property owner to transition from a construction loan to a less expensive permanent loan. Liens should be avoided at all costs. They are expensive to deal with, they cause bad feelings, and all parties involved receive bad press from them.

The project manager should work to protect the owner from liens from subcontractors and suppliers and from their subcontractors and suppliers. The most common liens are filed from third- or even fourth-tier parties. These may be from a supplier to a supplier to a subcontractor to a subcontractor. The farther removed, the more difficult it is to prevent and ultimately resolve.

The supplier's rights are protected by law to ensure payment for goods received by the end user and the ultimate beneficiary. In most jurisdictions, suppliers are required to file what is known as a materialman's notice, similar to the one illustrated in Box 11.2. This notice is sent to the property

## Box 11.2    Materialman's notice

# ATB Asphalt
### 518 Third Avenue
### Seattle, WA 99801

September 21, 2016

Green Development Company
1001 First Avenue
Seattle, WA 98001

Re:  Materialman's/Supplier's Notice to Owner

Dear Sir:

We are pleased to communicate that on September 20, 2016, our firm commenced to furnish materials for use on your construction project known as First and Pine Office Building, which is located at the premises commonly described as 1001 Pine Avenue NE, Seattle, WA 98101. Delivery was commenced at the request and upon the order of Northwest Construction Company, a contractor whose business address is 1242 First Avenue, Cascade, Washington 98202. Even though we do not have the opportunity of dealing with you directly, be assured that we are pleased to be able to make a contribution to your project.

In order to preserve our lien rights, the laws of this state require that we advise you that we are furnishing materials for use upon your property, and that we may claim a lien for the value of those materials.  We certainly have no reason to anticipate the necessity of making such a claim of lien, and trust that you will not construe this notification as any reflection upon either you or Northwest Construction Company.

Our experience has been that everyone in the construction industry recognizes that it is simply good business practice to protect one's lien rights.  We sincerely hope that you will be pleased with the materials we are furnishing, and trust that we may be of further service to you in the future.

Sincerely,

*Randy Jones*

Randy Jones
President

owner notifying them that the supplier will be delivering material to their project. The notice is technically required in order to preserve the supplier's lien rights. Some notices are also sent to the general contractor. The project manager should file all material notices received with each subcontract. At the end of the project, it is appropriate to request an unconditional lien release from each party who filed such a notice. If they went to the trouble of preserving their lien rights, they have the responsibility to release them when paid.

Conditional lien releases should accompany each request for payment from suppliers and subcontractors. The contract may require submission of interim (or conditional) lien releases from suppliers and subcontractors with each payment request to the owner. A project manager should require suppliers and subcontractors to submit conditional lien releases with their monthly pay requests, regardless of the owner's requirements. Interim lien releases should be collected and filed just in case a problem occurs. If the owner requests copies of them, the project manager should not resist, but comply. This also applies in both lump sum and GMP scenarios.

A conditional lien release is just that: The release of lien rights is conditional upon receipt of payment. An unconditional lien release means that all payments have been received and all lien rights are unconditionally released. Some payment request procedures will require conditional releases to accompany each request for payment along with an unconditional release for the previous month's payment. Some owners will go so far as to physically trade checks for lien releases.

There are several different release forms. Many states produce forms that comply with local law. Some owners will try to get project managers to release lien rights that are not required to be released by law. Releasing rights to claim for extra work are also sometimes hidden within the lien release or payment request forms. An example lien release form is shown in Box 11.3. Lien releases are an area where legal counsel should be sought. The conditional release covers the month for which payment is requested, and the unconditional release is for the month that was covered by the payment that was received. A final lien release will be submitted for the entire contract at close-out to receive retention release. This is also discussed in Chapter 17.

So what happens if a supplier does not get paid? The time limits involved vary with jurisdiction. Usually a pre-lien notice is filed; it is not a lien but just a warning that if the supplier does not get paid by a certain date, it will file a lien. Common timing is that if the supplier has not filed its lien within 90 days after delivery, the supplier has waived its lien rights. After the lien is filed, it has an additional 90 days to foreclose. Sometimes a lien is filed too quickly by a supplier. Liens are unfortunately (for the owner) easy to file but very difficult to have removed, and every precaution should be made to ensure they are not filed erroneously.

The project manager needs to follow all the appropriate rules and take all the necessary precautions but should not count on rules regarding timing of filing documentation for protection. If a supplier has delivered material for which the property owner is benefiting and the supplier has not been paid, some courts may find that the supplier is due payment, regardless of the technicalities with respect to filing.

Since liens are filed against the owner's property, why should the project manager worry about them? Liens are a problem. They reflect poorly upon all the parties involved. Liens are seldom filed by the GC. It is the owner who has received a lien from the contractor's supplier. The project manager is expected to protect the owner from these unfortunate occurrences. On the other hand, the contractor's right to lien and the lien itself are sometimes necessary construction management tools which the project manager must understand and use.

## Box 11.3    Lien release

### INTERIM LIEN/CLAIM WAIVER

From:  Northwest Construction Company
       1242 Cascade Ave
       Seattle, WA 98202

Project: NanoEngineering Building
       1101 NE 15th Ave
       Seattle, WA 98011

Contact Person: Robert Smith
Contact Telephone: (206)340-6653

Project Manager: Ted Jones
Telephone: (206) 239-1422

### CONDITIONAL RELEASE

The undersigned does hereby acknowledge that upon receipt by the undersigned of a check from *the University of Washington* in the sum of *$4,726,659.00* and when the check has been properly endorsed and has been paid by the bank upon which it was drawn, this document shall become effective to release any and all claims and rights of lien which the undersigned has on the above references project for labor, services, equipment, materials furnished and/or any claims through *April 30, 2016,* except it does not cover any retention or items furnished thereafter. Before any recipient of these documents relies on it, said party should verify evidence of payment to the undersigned.

### UNCONDITIONAL RELEASE

The undersigned does hereby acknowledge that the undersigned has been paid and has received progress payments in the sum of *$22,778,285.00* for labor, services, equipment, or materials furnished to the above referenced project and does hereby release any and all claims and rights of lien which the undersigned has on the above referenced project. The release covers all payment for labor, services, equipment and materials furnished, and/or claims to the above referenced project through *March 31, 2016* only and does not cover any retention or items furnished after that date.

I CERTIFY UNDER PENALTY OF PERJURY UNDER THE LAWS OF THE STATE OF WASHINGTON THAT THE ABOVE IS A TRUE AND CORRECT STATEMENT.

Signature:    ***Sam Peters***
            Vice President
            (Authorized Corporate Officer)

Date:    *July 31, 2015*

## Summary

Receipt of timely payment is one of the most important responsibilities of the project manager. The exact format for submitting payment requests will vary depending on the type of contract. An SOV is used to support payment applications on lump sum contracts and fee payments on cost-plus contracts. The project manager is responsible for developimg the payment request, making sure payment is received, and subsequently seeing that the subcontractors and suppliers are paid. If payment has not been received on time, the project manager should contact the owner to determine the cause. The financial relationship with the owner and the project are the project manager's responsibility. The same scenario holds true with respect to subcontractors and suppliers. The project manager ensures that they are paid promptly. Owners may withhold a portion of each payment to ensure timely completion of the project. This is known as retention. The retention rate is specified in the contract. Liens can be placed on a project if subcontractors or suppliers are not paid for their labor or materials. To preclude liens, owners require lien releases with payment applications.

## Review questions

1.  How does the type of contract influence the format of the payment request?
2.  What is an SOV? Why is it required by project owners?
3.  What type of data is submitted to support a payment request on a negotiated open-book contract?
4.  What type of data is submitted to support a payment request on a lump sum contract?
5.  What is retention? What is it used for?
6.  What is a lien?
7.  What is the relationship between a payment bond (Chapter 2) and a lien?
8.  What are interim lien releases used for?
9.  Why are final lien releases requested by owners as part of the final payment process?
10. How can slow payments from the owner affect the financial stability of a contractor?

## Exercises

1.  Prepare an SOV similar to the one shown in Worksheet 11.3, except for another sample case study project, such as a middle school, hotel, or office building.
2.  Prepare a pay request for the month ending January 2016 (or any other chosen month) for the NanoEngineering Building, including all supporting documentation and lien releases. Assume the project is on schedule.
3.  We do not advocate this but, for comparison, prepare a lump sum SOV for the NanoEngineering Building example except aggressively front load the general conditions and fee and other markups. How would this change the cash flow curves presented previously in Chapter 4?

# 12 Cost and time control

## Introduction

The project manager and superintendent are responsible for ensuring that a project is completed within the time allowed by the contract and within the project budget. To accomplish this challenging task, cost and time controls are established to monitor progress throughout the duration of the project. The project team wants to ensure that all contractual requirements are completed as early as possible while earning a profit on the project. Early completion of a project results in reduced jobsite overhead costs enhancing anticipated profits. In this chapter, we will examine several techniques for cost and time control.

## Cost control

Cost control begins with assigning cost codes to the elements of work identified during the work breakdown phase of developing the cost estimate. Work breakdown structure was discussed in Chapters 3 and 4. These cost codes allow the project manager and superintendent to monitor project costs and compare them to the estimated costs. The objective is not that the team has to rigidly keep the cost of each element of work under its estimated value but to ensure that the total cost of the completed project is under the estimated cost. This is analogous to driving on the road. You don't have to exactly follow the "line-down-the-middle-of-the-road," you can veer a little left and a little right, but you just have to stay on the road.

Some uses of actual cost data are:

- to monitor project costs, identify any problem areas, and select mitigation measures;
- to identify additional costs incurred as a result of changes in the project scope of work;
- to identify costs for completing work that was the responsibility of a subcontractor (called *back charges*);
- to develop a database of historical cost data that can be used in estimating the cost of future projects;

- to evaluate the effectiveness of the project management team; and
- to provide the project owner with a cost report, which may be an open-book contract requirement.

The cost control process involves the following steps:

- Cost codes are assigned to each element of work in the cost estimate.
- The cost estimate is corrected based on buyout costs (project buyout discussed in Chapter 8).
- Actual costs are tracked for each work item using the assigned cost codes.
- The construction process is adjusted, if necessary, to reduce cost overruns.
- Actual quantities, costs, and productivity rates are recorded and an as-built estimate is prepared.

While all costs should be monitored, the items that generally involve the greatest risk are:

- direct labor,
- equipment usage or rental, and
- jobsite overhead or project administration.

It is possible to lose money on material purchases but, with good estimating skills, it is not probable, and the risk is not as great as it is on labor. The same holds true with subcontractors. They have quoted prices for specific scopes of work, and the subcontractors, therefore, bear the risk associated with labor and equipment.

It is difficult to control costs if the project manager does not start with a detailed estimate. For example, let's suppose there was an $8,000 cost overrun on 4,000 square feet of concrete slab on grade. The aggregate cost analysis shown in Table 12.1 does not provide sufficient detail to identify the cause. Project managers should use a more detailed cost breakdown, as shown in Table 12.2, to determine the cause of the cost overrun.

**Table 12.1**  Aggregate cost analysis

|  | Quantity | Estimated unit price | Estimated cost | Actual unit price | Actual cost | Variance |
|---|---|---|---|---|---|---|
| SOG | 4,000 SF | $4/SF | $16,000 | $6/SF | $24,000 | +$8,000 |

**Table 12.2**  Detailed cost analysis

|  | Quantity | Estimated unit price | Estimated cost | Actual unit price | Actual cost | Variance |
|---|---|---|---|---|---|---|
| Formwork: | 440 LF | $4.10/LF | $1,800 | $4.30/LF | $1,900 | +$100 |
| Reinf. Steel: | 2 ton | $1,500/ton | $3,000 | $3,000/ton | $6,000 | +$3,000 |
| Concrete: | 50 CY | $120/CY | $6,000 | $140/CY | $7,000 | +$1,000 |
| Placement: | 50 CY | $24/CY | $1,200 | $22/CY | $1,100 | −$100 |
| Finish: | 4,000 SF | $1.00/SF | $4,000 | $2/SF | $8,000 | +$4,000 |
| Total system: |  | $4.00/SF | $16,000 | $8/SF | $24,000 | +$8,000 |

Now it is easy see that the problem is not with the carpenters forming the slab nor with the laborers placing the slab, but the bulk of the overrun is with reinforcing steel and concrete finishing. Why did this happen? Maybe the slab was rained on. Maybe the steel was not fabricated correctly. Maybe personnel changes were necessary. Maybe the estimate was too low. There could be a variety of reasons. The point is the project manager and the superintendent can now focus on evaluating these specific issues.

## Cost codes

To be able to control costs, they must be tracked accurately and compared against the corrected estimate. The first step is to record the actual costs incurred and input the information into a cost control database. Cost codes are used to allow comparison of actual cost data with the estimated values. There are several types of cost codes used in the industry. The best system to use on most projects is the coding system selected for the project files. Filing systems were introduced in Chapter 9. Many construction firms have adopted the CSI MasterFormat system described in Chapter 2. An example of this type of coding system is:

_____ . _____ . _____

(project number) (CSI work package) (element of cost)

The project number is assigned by the construction firm, the CSI work package code is from the MasterFormat, and the element of expense is the type of cost. An example element of cost coding is:

1. Labor.
2. Equipment.
3. Material.
4. Subcontract.

Using this system, the cost code for rough carpentry direct labor on the NanoEngineering Building would be:

9821.06-10-00.1

The cost code for rough carpentry material would therefore be:

9821.06-10-00.3

Depending upon the size of the construction firm, the type of work, and the type of owner and contract agreement, the project manager may perform job cost accounting in the home office or in the field. Generally, the smaller the firm and the smaller the contract value, the more likely all accounting functions will be performed in the home office. On larger projects, the project team may have a jobsite accountant. The type of contract and how it addresses reimbursable costs may also have some affect on where the construction firm performs the accounting function. Let's say, for example, the project is a $100 million pharmaceutical facility that has a negotiated guaranteed maximum price contract that allows for all on-site accounting to be reimbursed. It may be more cost-effective to perform accounting out of the home office with the assistance of an accounting department, but according to the terms of the contract, the owner will not pay for activities conducted off the project site.

Regardless of where the cost data are collected and where the checks are prepared, most of the accounting functions on a project are the same. The process begins with a corrected estimate. Then actual costs are incurred, either in the form of direct labor, material purchase, or subcontract invoice. Cost codes (those matching the file system and the estimate) are recorded on the time sheets and invoices. Often this process begins with the project engineer. The coded time sheets and invoices are then passed to the superintendent and the project manager for approval. Sometimes the officer-in-charge or maybe the client (on cost-plus projects) may also want to initial approval on each invoice. After the time sheets and invoices are coded and approved, the cost data are input into the cost control system.

One important aspect of this phase of cost recording is the accurate coding of actual costs. If costs are accidentally or intentionally coded incorrectly, the project team will not really know how they are doing on that specific item of work. Some superintendents and project managers may intentionally code costs against items where there is money remaining, not necessarily against the correct work activity, thereby hiding overruns. Others have been known to intentionally exceed the estimate on a specific line item of work to prove a point that the home office estimator is not covering the item correctly. Regardless of the reason, the project manager will not be able to monitor and collect accurate cost data if coding errors occur. All costs should be coded correctly to provide the project team with an accurate accounting of all expenditures.

## Work packages

Control or management of direct craft labor and equipment rental costs is the responsibility of the superintendent. The key to getting the field supervisors involved in cost control is to get their personal commitment to the process. One successful way for the project manager to do this is to have the superintendent actively involved in developing the original estimate. If the superintendent said it will take four carpenters working 2 weeks to form a particular cast-in-place concrete wall, he or she will endeavor to see to it that the task is completed within that time.

Work packages are a method of breaking down the estimate into distinct packages or assemblies or systems that match measurable work activities. These were previously discussed as part of the work breakdown in Chapters 3 and 4. For example, footings, including forming, reinforcing steel, and concrete placement, could be a work package. The work is planned according to the amount of hours in the estimate and monitored for feedback. When the footings are complete, the project manager and superintendent will have immediate cost control feedback. Work packages apply best to those who estimate and track costs by system rather than the pure CSI approach. When a system is complete, such as footings or slab on grade, the project manager and the superintendent immediately know how they are doing with respect to the overall estimate.

Some construction executives believe the field supervisors should not be told the budgeted value of each work package. This is a poor practice, because the field supervisors are key members of the project team and have critical roles in making the project a success. They should be provided with the actual budgeted cost, both in materials and man hours for each work package.

A good technique for monitoring project cost is to develop a direct work project labor curve similar to the one shown in Figure 12.1. It is important to have the superintendent or foreman record the actual hours incurred weekly and chart them against the estimate. If the actual labor used is under the curve, the field supervisors are either beating the estimate or behind schedule.

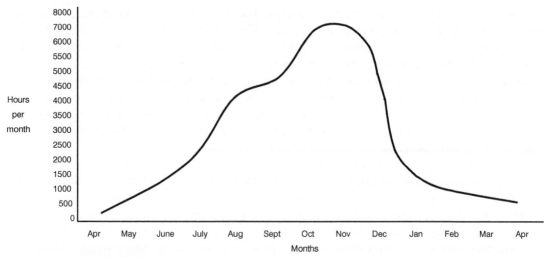

**Figure 12.1**   Project labor curve

The opposite is true if the hours are above the curve. This simple method of recording the man-hours provides immediate and positive feedback to the project team. It is better to use hours and not dollars when monitoring direct labor. The advantages of estimating using unit man-hours (UMH) over unit prices for labor were previously discussed in Chapter 3. Field supervisors think in terms of crew size and duration. They do not think in terms of $3.50 per square foot.

Which items should the project team track? The 80–20 rule applies here also: 80 percent of the cost and the risk lie with 20 percent of the activities, and it is those activities that deserve the most cost control attention. The project manager should evaluate and mitigate the risks. The estimate should be reviewed to identify those items that have the most labor hours or, in the case of equipment rental, the most cost. Work packages should be prepared for those items the project manager and superintendent believe are worthy of tracking and monitoring. Each work package should be developed by the foreman who is responsible for accomplishing the work. Figure 12.2 shows an example work package cost control worksheet.

## Management reports

The project manager is responsible for developing a monthly forecast for the project that will be shared with the superintendent and the officer-in-charge. The contractor's bonding and banking associates may also have an interest in the monthly forecast. The client may be included in the case of a cost-plus contract. This forecast includes line items for all areas of the estimate, cost to date, and estimated cost to complete. Each of the major areas of work receives a separate forecast page, and each of those is broken down for all categories of work. As outlined in Chapter 3, the major categories of the estimate include:

* direct labor,
* direct material,
* subcontractors and major material suppliers,

System/Area: Middle School
Foundations

October:

| Work Days: | 5 | 6 | 7 | 8 | 9 | 12 | 13 | 14 | 15 | 16 | 19 | 20 | 21 | 22 | 23 | 26 | 27 | 28 | 29 | 30 | 11/2 | Totals: |
|---|---|---|---|---|---|---|---|---|---|---|---|---|---|---|---|---|---|---|---|---|---|---|
| Estimated Crew Size: | 3 | 8 | 8 | 8 | 8 | 8 | 8 | 0.5 | 0 | 0 | 0 | 2 | 1.5 | 0.5 | 0 | 0 | 0 | 2 | 2 | 1 | 1.5 | |
| Estimated Daily Hours: | 24 | 64 | 64 | 64 | 64 | 64 | 64 | 4 | 0 | 0 | 0 | 16 | 12 | 4 | 0 | 0 | 0 | 16 | 16 | 8 | 12 | 496 |
| Accumulated Hours: | 24 | 88 | 152 | 216 | 280 | 344 | 408 | 412 | 412 | 412 | 412 | 428 | 440 | 444 | 444 | 444 | 444 | 460 | 476 | 484 | 496 | 496 |
| Actual Crew Size: | 0 | 4 | 6 | 8 | 10 | 10 | 10 | 10 | 1 | 0.5 | 0.5 | 0 | 0 | 0 | 0 | 0 | 2 | 2 | 0 | 0 | 0 | |
| Actual Daily Hours | 0 | 32 | 48 | 64 | 80 | 80 | 80 | 80 | 8 | 4 | 4 | 0 | 0 | 0 | 0 | 0 | 16 | 16 | 0 | 0 | 0 | 512 |
| Accumulated Hours: | 0 | 32 | 80 | 144 | 224 | 304 | 384 | 464 | 472 | 476 | 480 | 480 | 480 | 480 | 480 | 480 | 496 | 512 | 512 | 512 | 512 | 512 |

Manpower Curve:

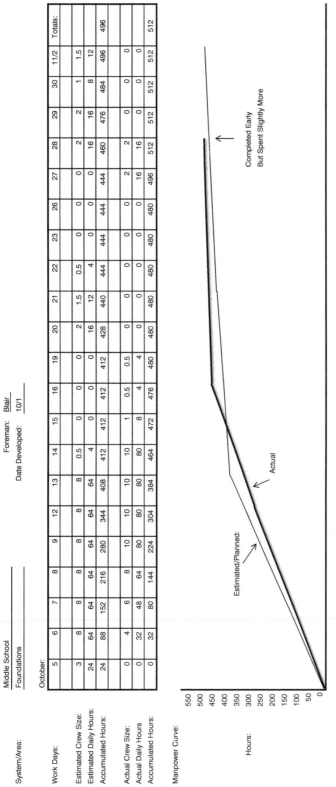

**Figure 12.2** Foreman's work package

- jobsite administration or general conditions, including equipment rental, and
- percentage markups including fee, excise tax, contingency, and insurance.

Worksheet 12.1 illustrates the project manager's summary forecast page for the NanoEngineering Building. It is a good practice to include a narrative with the monthly forecast explaining the significant differences from the previous month's forecast as well as a work plan for continuing or improving performance throughout the remainder of the project. The project manager cannot afford to wait until the project is finished to measure and report the overall project cost. It is not only too late to take corrective action; it is also too late to accurately determine why the team deviated from the plan.

There can be several other management reports either generated by the project manager or the accounting department according to the construction firm's practices and the requirements of any specific client or project. Most of the reports are computer generated and are accurate to the degree that the information regarding actual costs was accurately input. They can occur weekly or monthly, but either way, home office-generated cost control reports are likely to be too late for implementing any corrective action in the field.

## As-built estimate

Project managers often are too busy and excited about starting the next project to develop as-built estimates for a completed project. Accurate as-built estimates, like as-built drawings and as-built schedules, are important historical reporting tools. Revising the estimate with actual cost data assists in developing better future estimates. As-built estimates are part of the close-out process as will be discussed in Chapter 17. This is particularly true if actual unit price data, which require input of actual quantities as well as actual cost, are determined. These data can help with developing the firm's database as well as providing project managers and estimators with their own accurate cost factors for future use. There are many different types of databases that can be used for a variety of different purposes. There are several software systems available that allow a contractor to prepare a customized database.

## Schedule control

Schedule control involves monitoring the progress of each activity in the construction schedule and determining the impact of any delayed activities on the overall completion of the project. Schedule control is just as important as cost control, because the project team wants to ensure substantial completion is achieved prior to the required contractual completion date. If this does not occur, liquidated damages may be owed to the owner to compensate for the project's not being completed on time.

Schedule development and updates were discussed in Chapter 4. The superintendent should report construction progress at the weekly project coordination meetings. The project manager should review the progress with the project team and identify the causes of any delays. He or she should select appropriate mitigation measures with the superintendent. Such measures may

# Worksheet 12.1  Monthly forecast

Project: *NanoEngineering Building*      Job No: 9821      PM: *Jones*      Date: 4/15/2016

## Monthly Forecast

| CSI Division | Description | Estimate Totals | Change Orders | Current Contract | Cost To-Date | Cost To-Go | Forecast Cost | Variable +/- |
|---|---|---|---|---|---|---|---|---|
| 1 | General Conditions | $5,450,000 | $0 | $5,450,000 | $2,900,045 | $2,490,000 | $5,390,045 | $59,955 |
| 2 | Demolition | $850,000 | $0 | $850,000 | $835,000 | $0 | $835,000 | $15,000 |
| 3 | Concrete | $8,055,000 | $138,100 | $8,193,100 | $7,998,875 | $20,000 | $8,018,875 | $174,225 |
| 4 | Masonry | $1,525,100 | $0 | $1,525,100 | $0 | $1,525,100 | $1,525,100 | $0 |
| 5 | Structural & Misc. Metals | $803,030 | $0 | $803,030 | $458,000 | $340,000 | $798,000 | $5,030 |
| 6 | Carpentry Subtotal | $937,500 | $0 | $937,500 | $600,000 | $305,000 | $905,000 | $32,500 |
| 7 | Thermal & Moisture | $3,080,980 | $0 | $3,080,980 | $575,000 | $2,500,000 | $3,075,000 | $5,980 |
| 8 | Doors & Glass | $2,673,000 | $0 | $2,673,000 | $0 | $2,673,000 | $2,673,000 | $0 |
| 9 | Finishes | $4,795,350 | $0 | $4,795,350 | $0 | $4,795,350 | $4,795,350 | $0 |
| 10 | Specialties | $388,290 | $0 | $388,290 | $0 | $388,290 | $388,290 | $0 |
| 11-13 | Equipment | $2,740,350 | $0 | $2,740,350 | $0 | $2,740,350 | $2,740,350 | $0 |
| 14 | Elevator | $842,000 | $0 | $842,000 | $200,000 | $642,000 | $842,000 | $0 |
| 21-23 | Mechanical Systems | $15,322,923 | $0 | $15,322,923 | $2,450,000 | $12,750,000 | $15,200,000 | $122,923 |
| 26, 27 | Electrical Systems | $6,148,000 | $0 | $6,148,000 | $1,500,000 | $4,758,000 | $6,258,000 | –$110,000 |
| 31-33 | Sitework | $4,149,714 | $18,950 | $4,168,664 | $2,950,000 | $1,239,750 | $4,189,750 | –$21,086 |
| 90 | Tax, Insurance, Contingency | $1,485,365 | $4,080 | $1,489,445 | $0 | $1,489,445 | $1,489,445 | $0 |
| | **Total Cost:** | $59,246,602 | $161,130 | $59,407,732 | $20,466,920 | $38,656,285 | $59,123,205 | **$284,527** |
| | **Fee:** | $1,777,398 | $4,850 | $1,782,248 | | | | **Saving** |
| | **Total Contract:** | $61,024,000 | $165,980 | **$61,189,980** | | | | |

| | | |
|---|---|---|
| Forecast Loss: | | $0 |
| Forecast Savings: | | $284,527 |
| GMP Savings Split: | Client  70% | $199,169 |
| | NWCC  30% | $59,751 |

| | |
|---|---|
| Current Contract Fee: | $1,782,248 |
| **Forecast Final Fee:** | **$1,841,999** |

include expediting material delivery, increasing the size of the workforce, or working extended hours. Short-interval schedules would be developed to intensively manage all accelerated work.

Just as an as-built estimate is important for estimating future work, an as-built schedule is helpful in scheduling future projects. Similar to estimating, the greatest risk in scheduling is the determination of the direct craft workforce productivity. The project manager and home-office scheduler should develop a personal set of productivity factors based on actual prior experience. These factors will help the project manager and superintendent to establish realistic activity durations when scheduling future projects.

# Earned value

Earned value is a technique for determining the estimated or budgeted value of the work completed to date (or earned value) and comparing it with the actual cost of work completed. Before finding its way to the construction industry, earned value management first emerged in the 1960s as a financial analysis tool for the United States government. The most effective use of earned value to a general contractor's project manager is to track direct labor, which represents the construction team's greatest project risk. As discussed earlier in this chapter, man-hours generally are used for monitoring direct labor.

The work-package curve shown in Worksheet 12.1 shows the man-hours planned by the foreman and the actual man-hours used for the footing installation. During the first week when the actual man-hour curve was below the estimated curve, did this necessarily represent that the foreman was under budget? Could he have been behind schedule? The project team could actually be ahead or behind schedule and over or under budget and just about any combination between.

Introduction of a third, or earned value curve shows the estimated hours the foreman has earned based upon the estimated quantities that were actually installed. This curve is determined by plotting the total number of man-hours estimated for the work package multiplied by the cumulative percent completed. With this third curve, an actual measure of productivity can be made. This method of monitoring will provide more accurate feedback to the project team for appropriate correction.

The schedule status is determined by subtracting the actual time used to perform the work from the time scheduled for the work that has been performed. This is the same as measuring the horizontal distance between the earned value and the estimated curves shown in Figure 12.3. The actual time used to complete the budgeted or estimated 90 man-hours of work was 6 days, while the time scheduled for that amount of work was 9 days. Therefore, the foreman is 9 minus 6 or 3 days ahead of schedule.

The cost status is determined by subtracting the actual cost of work performed from the earned value of the work performed. This is the same as measuring the vertical distance between the actual and earned value curves. Looking again at Figure 12.3, we see that the actual value of work performed was 48 man-hours while the earned value was 90 man-hours. Therefore, the foreman is 90 minus 48 or 42 man-hours under budget. Figure 12.4 shows a different situation. The foreman is 48 minus 72 or 24 man-hours over budget and approximately 5 minus 6 or a little more than 1 day behind schedule.

| Work Days | 1 | 2 | 3 | 4 | 5 | 6 | 7 | 8 | 9 | 10 | 11 | 12 | Totals |
|---|---|---|---|---|---|---|---|---|---|---|---|---|---|
| Estimated Hours/Day | 10 | 10 | 10 | 10 | 10 | 10 | 10 | 10 | 10 | 10 | 10 | 10 | |
| Cumulative Hours | 10 | 20 | 30 | 40 | 50 | 60 | 70 | 80 | 90 | 100 | 110 | 120 | 120 |
| | | | | | | | | | | | | | |
| Actual Hours/Day | 8 | 8 | 8 | 8 | 8 | 8 | | | | | | | |
| Cumulative Hours | 8 | 16 | 24 | 32 | 40 | 48 | | | | | | | |
| | | | | | | | | | | | | | |
| Earned Hours/Day | 15 | 15 | 15 | 15 | 15 | 15 | | | | | | | |
| Cumulative Hours | 15 | 30 | 45 | 60 | 75 | 90 | | | | | | | |

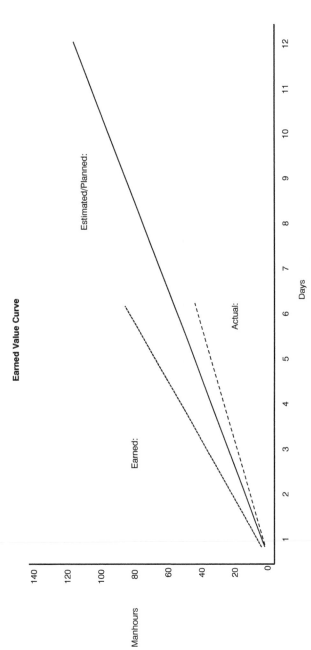

**Earned Value Curve**

**Figure 12.3**  Earned value curve

| Work Days | 1 | 2 | 3 | 4 | 5 | 6 | 7 | 8 | 9 | 10 | 11 | 12 | Totals |
|---|---|---|---|---|---|---|---|---|---|---|---|---|---|
| Estimated Hours/Day | 10 | 10 | 10 | 10 | 10 | 10 | 10 | 10 | 10 | 10 | 10 | 10 |  |
| Cumulative Hours | 10 | 20 | 30 | 40 | 50 | 60 | 70 | 80 | 90 | 100 | 110 | 120 | 120 |
| | | | | | | | | | | | | | |
| Actual Hours/Day | 12 | 12 | 12 | 12 | 12 | 12 | | | | | | | |
| Cumulative Hours | 12 | 24 | 36 | 48 | 60 | 72 | | | | | | | |
| | | | | | | | | | | | | | |
| Earned Hours/Day | 8 | 8 | 8 | 8 | 8 | 8 | | | | | | | |
| Cumulative Hours | 8 | 16 | 24 | 32 | 40 | 48 | | | | | | | |

**Earned Value Curve**

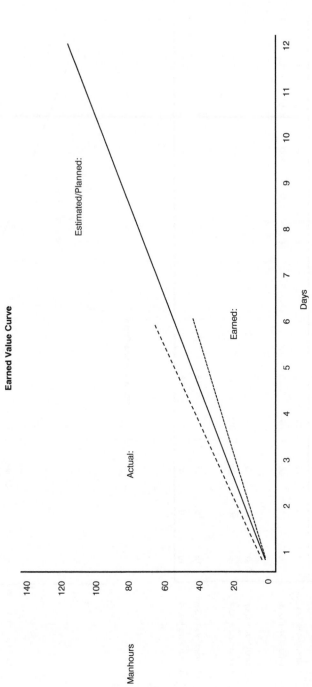

**Figure 12.4**   Earned value curve

## Summary

Project cost control and time control are essential project manager and superintendent functions. The project team wants to ensure that the project is completed within budget and within the time allowed by the contract. Cost codes are used to track project costs and compare them with the budgeted (or estimated) value for each element of work. Work package analysis provides the project manager and the superintendent a method for tracking direct labor cost that represents the greatest risk on the project. As-built estimates are developed to create unit price data that can be used on future estimates. Schedule control involves monitoring the progress of each scheduled activity and selecting appropriate mitigation measures to overcome the effects of any schedule delays. As-built schedules are prepared to allow project managers to develop historical productivity factors for use on future projects. Another technique for cost and time control is earned value analysis, which compares the budgeted value of work completed with the actual cost incurred.

## Review questions

1.  What are four uses of actual project cost data?
2.  Why is the original cost estimate corrected based on buyout data?
3.  What three types of project costs present the greatest risk to the project team?
4.  Why are cost codes used for monitoring actual project costs?
5.  What are project labor curves used for?
6.  What is a foreman's work package cost control work sheet used for?
7.  Other than the examples listed earlier, what three typical assemblies or systems would warrant development of a foreman's work package?
8.  How frequently should a project manager develop a forecast of the estimated cost to complete a project?
9.  What are as-built estimates used for?
10. What are as-built schedules used for?
11. What is an earned value analysis used for?
12. What would the cost codes be on the NanoEngineering Building for
    a.  Carpet?
    b.  Concrete placement?
    c.  Reinforcing steel purchase?

## Exercises

1.  Utilizing another case study project, prepare a work package for direct labor for a system of work such as concrete foundations, cast-in-place concrete walls, tilt-up concrete walls, or slab-on-grade. Attach the original portion of the estimate that refers to this work package. Assume the following percentages complete and spent:

| | Portion of work complete | Portion of estimate spent |
|---|---|---|
| Excavation: | 100% | 80% |
| Form work: | 90% | 100% |
| Reinforcement steel or mesh: | 85% | 110% |
| Concrete placement: | 50% | 40% |
| Slab finish (if appropriate): | 40% | 75% |
| Tilt-up panel erection (if appropriate): | 10% | 7% |

2. Prepare a cost forecast for the foregoing work activities. Based upon the work-packages developed, forecast the remaining hours needed and calculate the total over- or under-run this system will achieve. Convert the hours to dollars using wage rates from Chapter 3 or other local current rates. List out what possible reasons the over- or under-runs could be occurring and what corrective actions could be taken. If all continues on this same trend (proceeds at the same rate as it has been), calculate what the historical as-built UMH will be.

3. Prepare additional earned value curves starting with the footing work package example presented earlier in the chapter. Draw a curve where the foreman is over budget and behind schedule and another where the team is on budget and ahead of schedule. Prepare a narrative explaining why these situations might be occurring.

4. Prepare a list of the top ten systems or assemblies from the NanoEngineering Building which deserve the most cost control attention and fall within the 80–20 rule.

5. Referring to Figures 12.3 and 12.4 earned value curves, if each of these work packages proceeds at the same trend, when will they complete, and what will be the final hours over or under budget?

# 13 Quality management

## Introduction

Quality management is an important project management function. As stated in Chapter 1, it is one of the four critical attributes of project success, with the others being cost, time, and safety. It has short-term implications affecting material and labor costs on a project and long-term implications affecting the overall reputation of the construction firm. The firm's greatest marketing assets are satisfied owners, and delivering quality projects is critical to achieving customer satisfaction. The project manager and superintendent must work together to ensure that all materials used and all work performed on a project conform to the requirements of the contract plans and specifications. Nonconforming materials and work must be replaced at the contractor's cost, in terms of both time and money. This means that the contractor must bear the financial cost of tearing out and replacing the nonconforming work and that additional contract time is not granted for the impact the rework has on the overall construction schedule. Quality management is essential to ensure all contract requirements are achieved with a minimum of rework. This requires a proactive quality focus from all members of the project team.

Similar to project specific safety plans, as will be discussed in Chapter 14, quality control (QC) plans are developed during the preconstruction phase or project start-up to document the QC organization and procedures to be used on the project. Some owners on negotiated contracts require project QC plans to be submitted with cost and schedule proposals and project-specific safety plans during contract negotiation. Other owners may require that QC plans be submitted at the preconstruction meetings that have been discussed throughout this text.

Full-time QC inspectors often are specified in contracts for large, complex projects. Some contractors attempt to substitute a foreman or assistant superintendent who already has other responsibilities. If a full-time inspector is specified, the project manager must provide one and include the cost in the general conditions part of the estimate. This was the requirement on our case study project. On smaller projects, the QC and safety inspection functions may be combined and/or performed by supervisors with other responsibilities.

Quality management involves the following processes:

- setting quality standards,
- scheduling inspections,
- managing any required rework, and
- documenting corrections.

The exact procedures to be used and the organization of the QC team are identified in the project-specific QC plan. This is a prime example of "active" QC versus "passive" quality control: taking measures to prevent mistakes and ensure the project goes well the first time around compared to repairing bad work after the fact.

## Quality control planning

To achieve quality in the completed project, the project manager and superintendent must ensure systems are in place to ensure that quality materials are procured and received, quality craftsmen and subcontractors are selected, and all workmanship meets or exceeds contract requirements. This proactive approach to selecting effective procedures and processes to deliver quality projects is the focus of total quality management. This involves careful planning and an effective inspection process to ensure quality results. Variations from standards are measured, and actions are taken to correct all variations or deficiencies. Effective QC planning starts with a detailed study of the contract documents to determine all the project's unique quality and testing requirements. A project-specific QC plan is developed that identifies QC systems for each major work package. Qualified inspectors are selected and trained, if necessary, to ensure that all self-performed and subcontractor work meets contract specifications. These inspectors also should participate in the submittal process to ensure that quality materials are being submitted and that only approved materials are installed. Mock-ups, which are stand-alone samples of completed work, should be planned to establish workmanship standards for critical architectural features. Systems testing, balancing, and commissioning are scheduled to ensure all systems meet contractual performance criteria.

Checklists often are developed and included in the QC plan to ensure inspectors do not overlook critical items. An example QC checklist for concrete formwork is shown in Box 13.1. Formal QC reports should be required and acted upon. These reports list any deficiencies that have been identified as a part of the inspection process. All deficiencies identified should be listed on a deficiency list, nonconformance-report or QC log such as the one shown in Worksheet 13.1. The superintendent should discuss the status of uncorrected deficiencies during the weekly owner/architect/contractor coordination meeting, which was described in Chapter 9.

Proper materials must be procured to ensure conformance with contract requirements. Shop drawings and certificates of compliance with reference standards submitted by the suppliers and subcontractors must be carefully reviewed to ensure that all materials used on the project meet contract requirements. As discussed previously, the QC inspectors should actively participate in the submittal process so they can become familiar with the materials being proposed and have an opportunity to compare characteristics of submitted, and later delivered, materials with contractual requirements.

# Box 13.1    QC inspection checklist

### NORTHWEST CONSTRUCTION COMPANY
1242 First Avenue, Cascade, WA 98202

(206) 239-1422


### INSPECTION CHECKLIST
### 03-10-00 – Concrete Formwork

| | Conforms? |
|---|---|
| Date: _9/16/15_ | Yes    No |
| Project: _NanoEngineering Building_ | |

|  |  | Yes | No |
|---|---|---|---|
| 1. | Location, dimensions, and grades are as required: | X | |
| 2. | Formwork materials as specified: | X | |
| 3. | Re-used formwork is properly conditioned for re-use: | X | |
| 4. | Completed formwork provides structural sections: | X | |
| 5. | Temporary spreaders are removed: | X | |
| 6. | Forms secured against concrete pressure: | X | |
| 7. | Forms are aligned: | X | |
| 8. | Forms properly cleaned and treated: | Need wash down | |
| 9. | Shoring checked for location and bearing: | NA | |
| 10. | Free movement of expansion joints: | X | |
| 11. | Filler installed and secured: | X | |
| 12. | Ties are of proper size and spacing: | NA | |

**Notes:** _Inspector noted: 1) Dirt on rebar, 2) Rebar on ground, 3) Scrap wood in form. All these conditions were corrected the same day. Photographs are in the file._

Inspector: **_Jim Bates_**, _Project Superintendent_

# Subcontracted work

As discussed in Chapter 6, subcontractors perform most of the work on a construction project. The general contractor's reputation for quality work, therefore, is greatly affected by the quality of each subcontractor's work. Quality subcontractors are not necessarily the least-cost subcontractors but, in the long term, they will provide a superior project. The term _best-value subcontractors_ applies here as well.

# Worksheet 13.1   QC deficiency log

## PROJECT DEFICIENCY LOG

PROJECT: *NanoEngineering Building*
NUMBER: *9821*

SUPERINTENDENT: *Jim Bates*

PAGE: *1 of 1_PROJECT*

| No. | Description | Location | Responsible Party | Date Identified | Date Corrected | Date Re-Inspected |
|-----|-------------|----------|-------------------|-----------------|----------------|-------------------|
| 1 | Silt fence blew down this weekend | West and south | NWCC | 9/15 | 9/15 | 9/15 |
| 2 | Mud on foundation rebar | Grids 4 and 5 | Rebar Sub | 9/15 | 9/16 | 10/18 |
| 3 | Rebar stored on ground and wet | Shakeout yard | Rebar Sub | 9/16 | 9/16 | |
| 4 | Wood debris in footing forms | Footings F5 and F6 | NWCC | 9/16 | 9/16 | |
| 5 | Concrete break at 2250 PSI | Elevator pit | NWCC | 11/1 | | |
| 6 | EMT in SOG in lieu of PVC | SOG | Richardson | 11/1 | 11/5 | 11/7 |
| | | | | | | |

The project superintendent must ensure the jobsite is ready for subcontractors before scheduling them to start work. Quality requirements should be emphasized at subcontractor preconstruction meetings. Subcontractors may be required to prepare mock-ups for exterior or interior finishes to establish the required level of workmanship. They must fully understand their quality requirements before being allowed to start work on the project. Subcontractors should not be allowed to build over work that has been improperly installed by a previous subcontractor.

The superintendent should walk the project when each subcontractor finishes its portion of the work to identify any needed rework. Any deficiencies noted should be listed in the log shown in Worksheet 13.1. All rework should be completed before follow-on subcontractors are allowed to start work. Deficiencies should be corrected as the work progresses to minimize the size of the punch list at the pre-final inspection.

## Testing and inspection

QC starts first with diligent subcontractor selection and second with the submittal process. Approved materials and equipment are then procured using the methods described in Chapters 7 and 8. All materials and equipment should be inspected as they are being delivered to the project site. They should be examined and compared to the appropriate purchase orders and submittal documentation. If the materials and/or equipment are not as specified, they should not be unloaded from the truck and should be rejected. Materials stored on site also need to be inspected periodically to ensure they are secured and properly protected against weather damage.

Mock-ups required either by contract or the QC plan are constructed prior to initiating work to establish workmanship standards. An example of a mock-up is a separate, stand-alone, 6-foot-by-6-foot section of a brick wall to establish the brick pattern and grout workmanship. Mock-ups are examined as a part of the preparatory inspection for each phase of the work. These inspections should ensure that all preliminary work has been completed on the project site either by the general contractor's work force or the appropriate subcontractor's. All materials and equipment are inspected to ensure they are appropriate for the tasks to be performed.

Once the work starts, the inspector conducts an initial inspection to ensure workmanship meets quality standards and dimensional requirements are met. Daily follow-up inspections are conducted to ensure continuing compliance with contract requirements. QC inspectors prepare daily reports, similar to the one illustrated in Box 13.2, to document their inspection activities and any deficiencies identified. These daily inspection reports should be reviewed both by the superintendent and by the project manager. Outstanding deficiencies should be entered in the deficiency log and tracked until they have been corrected. Building inspectors and design team members should be scheduled to visit the site throughout the completion of the project to ensure that all work is inspected prior to being covered. This avoids the need to uncover work for inspection.

Article 3.12 of the ConsensusDocs 500 contract for the NanoEngineering Building addresses uncovering work for inspection. Work covered contrary to the architect's request or to contract requirements is to be uncovered at the contractor's expense. The architect may request that additional work be uncovered for inspection. If the uncovered work is found to conform to contract requirements, a change order will be issued to cover the impact costs. If the work is found not to conform to contract requirements, the inspection and correction costs are the contractor's responsibility.

## Box 13.2    QC daily inspection report

### NORTHWEST CONSTRUCTION COMPANY
1242 First Avenue, Cascade, Washington 98202
(206) 239-1422

### DAILY INSPECTION REPORT

Date:                    *September 16, 2015*

Project:                 *NanoEngineering Building*

Weather:                 *Rain and Gray and 10MPH SW Wind*

Work Inspected:          *Footing Forms prior to pour*

*Work found generally to be in accordance with approved plans and specifications with exceptions as noted below. I talked to Superintendent Bates and these items will be corrected prior to the concrete placement scheduled for this Friday.*

Deficiencies Noted:
1) *Some mud on the rebar*
2) *Some rebar laying on the ground, not yet hung, currently in a puddle since last week's rain storm*
3) *Minor wood and debris in footing forms requires removal prior to pour*

Inspector:  **Jeremy Shaker**, TTI Testing, Inc.

The specifications identify the items to be tested and how they are to be tested. Other testing directed by the architect or owner requires a change order to compensate the contractor for the additional work. Professional testing and inspection agencies are used to perform required testing. Contracts often require independent monitoring and testing of:

- underground utilities,
- soil compaction,
- shoring systems,
- concrete,
- reinforcing steel,
- post-tension cables,
- pre-cast concrete,

- structural steel fabrication, erection welds, and field bolt torque,
- fireproofing,
- roofing,
- waterproofing,
- glazing, windows, and curtain-walls,
- waterproofing, and
- elevators.

Reports provided by testing agencies are reviewed by the QC inspectors and filed with their daily inspection reports to the general contractor's superintendent. The contract may require that copies of the reports also be submitted to the owner. Materials not meeting specified standards must be replaced.

When the project is nearing completion, the project manager schedules the pre-final inspection. Generally, this inspection is conducted by the superintendent, an owner's representative, and a representative of the design firm. All deficiencies noted are recorded on a punch list, which is a list of items to be completed or corrected. If an active quality management system has been used on the project, the size of the punch list should be minimal. When all deficiencies listed on the punch list have been corrected, the project manager schedules the final inspection to verify that all required work has been completed in conformance with contract requirements. The final inspection is also conducted by the superintendent, an owner's representative, and a representative of the design firm. Punch list and other aspects of project close-out are discussed in Chapter 17.

## Summary

An effective quality management program is essential to project success, in terms of both a satisfied owner and a profitable project for the general contractor. Poor-quality work costs the contractor both time and money and can cause the loss of future projects from the owner. The project manager and superintendent must ensure that all materials and work conform to contract requirements. The objective of a quality management program is to achieve required quality standards with a minimum of re-work. Quality materials must be procured, and qualified craftsmen must be selected to install them. Workmanship must meet or exceed contractual requirements.

QC inspectors should be used to inspect all work, whether performed by the contractor's workforce or by a subcontractor's. Materials are inspected for conformance with contract specifications upon arrival on the jobsite. Mock-ups may be required to establish workmanship standards. Each phase of work is inspected as it progresses, and a pre-final inspection is conducted when most of the work has been completed. All deficiencies identified during the pre-final inspection are recorded on a punch list. When all outstanding deficiencies have been corrected, the final inspection is conducted.

## Review questions

1. What is the primary objective of active quality management programs?
2. What four processes are used in managing quality on a construction project?

3. Why are project-specific QC plans needed? Why are generic QC plans not acceptable?
4. What are mock-ups, and why are they used?
5. Why should QC inspectors be involved in the submittal process?
6. Why should subcontractors not be allowed to build over work that was improperly done by a previous subcontractor?
7. Why are materials that are stored on site periodically inspected?
8. What is the purpose of a pre-inspection?
9. What is the purpose of an initial inspection?
10. What is a pre-final inspection, and when is it conducted?

## Exercises

1. What QC organization do you recommend the contractor use on the NanoEngineering Building?
2. What are two mock-ups the general contractor should require subcontractors to construct for the NanoEngineering Building?
3. Develop a QC checklist for the concrete columns or another significant assembly for the NanoEngineering Building, making reference to the technical specifications.
4. Develop a QC checklist for gypsum board assemblies for the NanoEngineering Building, making reference to the technical specifications.
5. List five examples each of active and passive quality control.

# 14 | Safety management

## Introduction

Construction is one of the most dangerous occupations in the United States, accounting for about 10 percent of the disability injuries and 20 percent of the fatalities that occur in the industrialized work force. There are two major aspects of project site safety: (1) safety of persons working on the site and (2) safety of the general public who may be near the project site. Both aspects must be addressed when developing project-specific safety plans.

The primary causes of jobsite injuries are:

- falling from an elevation,
- being struck by something,
- trenching and excavation cave-in,
- being caught between two objects, and
- electrical shock.

Within the United States, the Occupational Safety and Health Administration (OSHA) has the primary responsibility for establishing safety standards and enforcing them through inspection of construction work sites. The Occupational Safety and Health Act that established OSHA contains a provision that allows states that desire to administer their own industrial safety programs to do so as long as their requirements are at least as stringent as those imposed by OSHA. About half of the states have opted to administer their own programs. In this text, we discuss only the OSHA requirements, but they are similar to those imposed by state-mandated programs.

Most construction companies have recognized the importance of safety management and have developed effective company safety programs that include new employee orientation, safety training, and jobsite safety surveillance. The effectiveness of these programs, however, is directly related to management's commitment to safety. Project managers and superintendents are responsible for the safety of workers, equipment, materials, and the general public on their project sites. They must set the standard regarding safety on their projects and enforce safety standards at all times. A continual safety awareness campaign is needed that is focused on reducing accidents. Frequent, at least weekly, safety inspections should be conducted of the jobsite to identify hazards and ensure compliance with job-specific safety rules. Every project meeting should address safety in some manner. Most construction firms require their superintendents or foremen to conduct weekly safety meetings with workers to maintain a continuous emphasis on hazard removal and safe work practices.

Many hazards exist on all construction sites, and mitigation measures are needed to minimize the potential for injury. Most construction projects are unique, and construction workers are constantly expected to familiarize themselves with new working environments. In addition, craft workers might work on a project site only during certain phases of work and then move on to another project site. This continuing change in the composition of the work force presents significant challenges to the project team. Another significant safety challenge is the increased employment of workers for whom English is a second language. These workers often have difficulty reading and understanding safety signage, but this does not release the general contractor from the responsibility to ensure that all personnel working on the jobsite do so in a safe manner.

Jobsite safety is a significant project management issue. Creating a safe working environment is a function of the physical conditions of the working environment and the behavior or working attitude of individuals working on the site. A safe working environment results in increased worker productivity and reduces the risk of injury. Accidents are costly, leading to disruption of the construction schedule, and require significant management time for investigation and reporting. The direct costs of accidents include:

- increased worker's compensation insurance premium cost,
- increased general liability insurance premium cost, and
- legal expenses associated with claim resolution.

Both worker's compensation insurance and general liability insurance were discussed in Chapter 2.

The indirect costs of accidents often are up to four times the direct costs and may include:

- first aid expenses,
- damage or destruction of materials,
- clean-up and repair costs,
- idle construction equipment cost,
- unproductive labor time,
- construction schedule delays,
- loss of trained labor,
- work slowdown,
- administrative and legal expenses,
- lowered employee morale, and
- third-party lawsuits.

OSHA compliance inspectors can issue citations for safety violations and shut down operations considered to be life-threatening. Stiff fines may be levied for violations.

Section 3.13 of the ConsensusDocs 500 Construction Manager-at-Risk contract available on this book's companion website addresses jobsite safety. Similar provisions should be included in all subcontracts. Section 5 of the ConsensusDocs 751 subcontract also on the companion website addresses jobsite safety.

## Accident prevention

Jobsite accidents occur as a result of either unsafe conditions or unsafe acts or both. Accident prevention requires a commitment to safety, proper equipment and construction procedures, regular jobsite inspection, and good planning. Everyone on the jobsite must understand the need to work safely and be alert for any unsafe conditions. Techniques that superintendents should use are to:

- involve everyone in the identification of potential hazards,
- involve everyone in the selection of procedures to eliminate or mitigate potential hazards,
- teach workers the proper use of personal protective equipment (PPE) and require that it be worn properly at all times,
- ensure that the jobsite is kept orderly and clean, and
- ensure that everyone understands proper procedures in the event of an emergency on the jobsite.

Jobsite accident prevention starts with an analysis of all the hazards that are associated with each task that is to be performed on the project. This job hazard analysis is performed during the project planning process that was discussed in Chapter 5. It involves the following three elements:

- a description of the construction operation,
- a listing of the hazards associated with the operation, and
- a plan for eliminating, reducing, or responding to the hazardous situations.

This job hazard analysis becomes the primary focus of the phased project-specific accident prevention plan that is discussed later in this chapter. A form such as the one illustrated in Box 14.1 can be used for the analysis.

Safe construction procedures and techniques should be identified for each phase of the work to minimize the potential for accidents. Some of the ways to reduce the risk created by a hazard are to:

- modify construction techniques to eliminate or minimize the hazard,
- guard the hazard, such as fencing the site,
- give a warning, such as back-up alarms on mobile equipment or warning signs,
- provide special training, and
- provide PPE.

PPE includes:

- body protection, such as protective clothing,

# Box 14.1    Job hazard analysis form

## NORTHWEST CONSTRUCTION COMPANY
*1242 First Avenue, Cascade, Washington 98202*
*(206) 239-1422*

### JOB HAZARD ANALYSIS

Project:  *NanoEngineering Building Project*

Phase: *Mass excavation*

| Sequence of Construction Tasks | Potential Hazards | Hazard Mitigation Measures |
|---|---|---|
| *Public safety* | *Automobile and pedestrian traffic* | 1)  *Mark site perimeter with fence, signs, and caution tapes*<br>2)  *Use full-time flagger wearing orange vest* |
| *Unloading equipment* | *Equipment sliding off trailer* | 1)  *Park trailer on firm solid surface*<br>2)  *Set equipment brake while unloading*<br>3)  *Use spotter to monitor vicinity* |
| *Site excavation* | *Struck by* | 1)  *Ground personnel wear high visibility vests*<br>2)  *Ground personnel stay outside of equipment swing radius* |
| *Site excavation* | *Noise* | 1)  *Equipment operators and ground personnel wear hearing protection* |
| *Loading earth into dump trucks* | *Struck by* | 1)  *Ensure path is clear before swinging bucket* |
| *Shutting down equipment* | *Unexpected equipment movement* | 1)  *Block wheels*  2)  *Set equipment brakes*<br>3) *Lower attachments to ground*<br>4)  *Lock equipment* |
| **Equipment to be Used** | **Inspection Requirements** | **Training Requirements** |
| *Track excavator Dump truck* | *Hydraulics, mirrors, and clean windshield* | *Inspect every vehicle entering and leaving the job site* |

Prepared by:  ***Jim Bates***                    Date: *March 30, 2015*

*CC: subcontractors, employees, architect, owner, project manager, and project engineer*

- eye protection, such as safety glasses and goggles,
- foot protection, such as safety shoes,
- hand protection, such as gloves,
- head protection, such as hard hats,
- hearing protection, such as earplugs and earmuffs,
- protection from falls, such as lifelines and safety harnesses, and
- respiratory protection, such as respirators.

Project managers and superintendents identify PPE requirements as part of the job hazard analysis and require that all general contractor and subcontractor workers use the proper equipment on the construction site. These requirements are specified in the phased project-specific accident prevention plan discussed later in this chapter. The wearing of hard hats, safety glasses, and safety shoes generally is required of all personnel at all times when working on the jobsite.

The hazards associated with each phase of work and selected mitigation strategies should be discussed with both general contractor and subcontractor work crews prior to allowing them to start work. Many superintendents require daily safety meetings prior to allowing the workers to start work. These meetings address risks and mitigation strategies for the work to be performed that day.

## Worker's compensation and injured worker management

All states have worker's compensation laws mandating coverage for all workers on a jobsite. Some states are monopolistic and provide worker's compensation insurance through state-administered funds, while others rely on private insurance companies. Worker's compensation is a no-fault insurance program that protects a contractor from being sued by their employees as a consequence of injuries sustained on the jobsite and provides compensation to workers who are injured or contract an illness on the job. Worker benefits include:

- coverage of medical expenses,
- compensation for lost wages,
- vocational rehabilitation, and
- disability pensions.

Worker's compensation premiums have two components. The first is the base premium that is applied to each $100 of direct employee compensation. The base premium is determined by the following equation by using the base rate published for each work classification. For example, the base rate for a carpenter may be $2.50 per $100 earned while the base rate for a roofer may be $6.70. Base rates are determined based on the degree of injury risk being assumed by a specific craft.

$$\text{Base premium} = \frac{(\text{direct wages})(\text{base rate})}{\$100}$$

The second component is the premium modifier, known as the experience modification rating, or EMR as it is known in the industry. It is based on the volume of claims made by a contractor's employees during the oldest 3 of the past 4 years. Typical company EMRs range from 0.5 to 2.0, with the average EMR being 1.0. Companies with good safety records typically have EMRs below 1.0.

The actual premium paid by a construction company is:

Modified premium = (base premium)(EMR)

For example, if the combined monthly base premium for a construction company were $70,000 and its EMR was 0.7, the actual premium cost would be (0.7)($70,000) or $49,000. However, if the company's EMR was 1.2, its actual premium would be (1.2)($70,000) or $84,000.

Since a contractor must pay worker's compensation premiums for each hour each worker is on the jobsite, they significantly add to the labor cost. Firms that have poor safety records have high EMRs, making their labor costs higher than those with good safety records and resulting low EMRs. Project managers and superintendents, therefore, must emphasize safety on their projects so their employers earn low EMRs and have competitive labor rates.

Injured workers need to be returned to the work force as soon as possible. Any worker injured on a project needs to be taken to a physician for examination and determination of the extent of injuries and the type of work, if any, that the injured worker is able to perform. Based on the physician's instructions, the construction company should devise a return-to-work strategy for the worker. This may include recuperation, physical therapy, reduced work hours, and/or alternative work assignment. Providing return-to-work options benefits the injured worker and reduces the workers' compensation claims cost that can adversely affect the company's EMR.

## Substance abuse

As was mentioned at the beginning of this chapter, construction is one of the most dangerous occupations in the United States. Workers under the influence of drugs or alcohol on the jobsite pose serious safety and health risks, not only to themselves but to all those who work in the proximity of the users.

To combat the problem of substance abuse, many contractors have established specific policies prohibiting substance use on project sites and people from working on site who are under the influence. A key element of this program is a requirement for periodic drug and/or alcohol testing. Testing requirements usually apply to all individuals working on the project site, whether they are employed by the contractor or by a subcontractor. Subcontractors generally are required to administer their own substance-testing programs, or they are not allowed to work on the project site. Individuals refusing testing or failing to pass a test are removed from the jobsite. Many contractors require testing during the hiring process, after an accident, and when a supervisor believes the individual's behavior warrants testing. Others also require random testing.

Contractors use third parties to administer their testing programs to ensure fairness in implementation and compliance with legal requirements. Selected administrators should be members of the Substance Abuse Program Administrators Association to ensure they keep updated on rules and court cases at state and national level.

## Hazardous materials communications

OSHA's Hazard Communication Standard requires chemical manufacturers, distributors, or importers to provide safety data sheets (SDSs, formerly known as *material safety data sheets*) to communicate the hazards of all hazardous chemical products. OSHA requires that all personnel working on a construction site be informed of all chemicals being used on the site and of any harmful effect they may cause. The transmittal of this information is to be accomplished by means of a hazardous materials communication program that includes:

- warning labels on all containers indicating the specific hazard posed,
- posting of SDSs, and
- specialized training regarding the safe handling, transporting, storage, and use of the chemicals.

SDSs must be obtained for each chemical used on site. They basically are short technical reports that identify all known hazards associated with particular materials and provide procedures for using, handling, and storing the materials safely. Information contained on these sheets includes:

- Section 1: identification;
- Section 2: hazard(s) identification;
- Section 3: composition/information on ingredients;
- Section 4: first-aid measures;
- Section 5: fire-fighting measures;
- Section 6: accidental release measures;
- Section 7: handling and storage;
- Section 8: exposure controls/personal protection;
- Section 9: physical and chemical properties;
- Section 10: stability and reactivity;
- Section 11: toxicological information;
- Section 12: ecological information;
- Section 13: disposal considerations;
- Section 14: transport information;
- Section 15: regulatory information; and
- Section 16: other information.

A copy of the SDS for each hazardous chemical must be made available to all potentially exposed workers before the chemical is used on the project site. Copies of all sheets for materials used on site should be maintained in an accessible location in the project office.

## Phased project-specific accident prevention plan

Phased safety planning is a systematic approach that examines the sequence of required construction activities and identifies the potential hazards associated with those activities. The planning process requires an understanding of construction practices and the site-specific conditions in which they will be performed. This list of potential hazards will guide the identification of activity-specific hazard mitigation measures. The technique for addressing these potential hazards is completion of job hazard analysis forms like the one shown in Box 14.1.

Some project owners require the general contractor to submit such plans at preconstruction conferences. Other owners may require that these plans be submitted with the proposal package on negotiated contracts. A phased project-specific accident prevention plan serves several purposes:

- identifies potential hazards that may occur on the jobsite during each phase of construction and identifies specific avoidance or mitigation measures to be taken;
- lists equipment to be used on the project, the hazards associated with each, and specific safeguards to be taken;
- lists construction methods to be employed during each phase of construction and safety procedures that are to be incorporated with each method; and
- describes the specific safety awareness training program to be used on the site.

The table of contents for the phased accident prevention plan for the NanoEngineering Building project is shown in Box 14.2. A job hazard analysis, such as illustrated in Box 14.1, was prepared for each phase and included in the plan.

## Reporting requirements

All construction firms employing more than ten individuals are required to maintain detailed records for OSHA. They are required to maintain a log of all jobsite injuries and illnesses occurring during the year and submit an annual report to OSHA no later than February 1 of the succeeding year. The log must contain the following information for each recorded illness or injury:

- the date of the illness or injury,
- the name of the affected employee,
- the occupation of the employee,
- a description of the illness or injury, and
- the amount of time lost due to the illness or injury.

In addition to the log, contractors are required to submit a separate one-page description of each illness or injury. This supplementary record of occupational illness or injury includes a description of how the illness exposure or injury occurred, the place of the exposure or accident, the employee's activity at time of exposure or injury, and the exact nature of the illness or injury to include affected body parts.

## Accident investigation

All accidents and near-miss incidents should be investigated promptly regardless of whether injuries and/or damage occurred. An inspection team may be needed for serious accidents, such as collapse of some portion of a project. A single investigator should be adequate for less-complex accidents. Supervisors or safety professionals should be used to conduct accident investigations to avoid biased findings. The objective of the investigation is to:

- determine what happened,

**Box 14.2   Phased accident prevention plan for NanoEngineering Building**

### Table of Contents

1. Project Information
2. Company Safety Policy
3. Project Administration
   a. Discipline and Accountability
   b. Safety Inspection
   c. Accident Investigation
   d. Record Keeping and Reporting
   e. Training
   f. Medical Assistance and First Aid
   g. Emergency Response Plan
4. Phased Project Planning: Site-Specific Hazards
   a. Public Safety
   b. Underground Utilities
   c. Mobilization
   d. Excavation and Foundation
   e. Structure Construction
   f. Exterior Enclosure
   g. Interior Enclosure
5. Technical Sections
   a. Excavation Safety
   b. Confined Space
   c. Concrete Forming and Placement
   d. Electrical Safety
   e. Scaffold Safety
   f. Fall Protection
   g. Ladder Safety
   h. Powered Aerial Work Platforms
   i. Tool Safety
   j. Hazardous Communications
   k. Fire Safety
   l. Housekeeping

## Box 14.3   Accident investigation form

### *NORTHWEST CONSTRUCTION COMPANY*
*1242 First Avenue, Cascade, Washington 98202*
*(206) 239-1422*

## Accident Investigation Form

Project Name: _____

Project Location: _____

1. Name of injured person: _____

2. Date and time of accident: _____

3. Job title of injured person: _____

4. Nature of injuries sustained by injured person: _____
_____

5. Describe the accident and how it occurred: _____
_____

6. What was the cause of the accident?: _____
_____

7. Were all company safety policies being followed? _____ If, no, explain: _____
_____

8. Was PPE required? _____ Was all required PPE being worn? _____If no, explain:
_____

9. Describe the job site conditions at the time of the accident: _____
_____

10. Had the injured person been properly trained to perform the assigned tasks? _____

11. Witnesses:_____

Prepared by _____ Date _____

- determine why the accident or incident happened, and
- determine procedures or policies that should be adopted to minimize the potential for future occurrence of such accidents or incidents.

The results of the investigation should be recorded by the superintendent and retained in the project files. A form similar to the one illustrated in Box 14.3 may be used to record the investigator's findings and recommendations. Photographs should be included with the report to document conditions at the accident site. Some owners will require a copy of each accident report. If injuries occur as a consequence of the accident, a report must be submitted to OSHA, and a copy must be kept on file with the construction firm for 5 years.

## Summary

Construction projects are dangerous work environments that pose many hazards to people working on them. To minimize the potential for worker injury, OSHA has established national safety standards that are enforced through a jobsite inspection program. Because of the importance of safety management, most construction firms have developed safety programs and hired safety professionals. Since project managers and superintendents are responsible for safety on their project sites, they must establish safety awareness programs and enforce good safety practices.

The key to a good safety program is accident prevention. This is accomplished by identifying all the hazards that are associated with each work activity and developing plans for eliminating, reducing, or responding to these hazards. This is known as *job hazard analysis*. Substance abuse also can result in accidents, so an effective substance abuse program is needed to remove from the jobsite individuals who are under the influence. PPE is provided, and its use is required to reduce the risk of jobsite injuries. Workers need to be informed of all hazardous chemicals that will be used on the project, their potential effects, and emergency and first aid procedures. Phased safety plans are developed for each project to identify hazards likely to occur during each phase of the work and specific measures to be taken to reduce the likelihood of accidents.

All accidents or near-miss incidents should be thoroughly investigated for lessons learned. The objective is to determine why the accident or incident occurred and identify procedures or policies to minimize the potential for future occurrence of similar accidents or incidents. A construction firm's safety record has significant impact on its labor cost. Worker's compensation premium rates are adjusted based on the history of claims submitted by the firm's employees.

## Review questions

1. What are the four prime causes of injuries on construction projects?
2. What is OSHA's role in safety management on construction sites?
3. Why is jobsite safety a significant project management issue?
4. What are four costs that can result from a construction site accident that typically are not covered by insurance?
5. What is the purpose of a job hazard analysis? What three elements are involved in the analysis?
6. What are three ways to reduce the risk created by a hazard on a jobsite?

7.  Why is substance abuse an important safety issue on a construction jobsite?
8.  How are substance abuse programs in the construction industry typically administered?
9.  What are four types of PPE that are commonly used on construction projects?
10. What is a hazardous materials communication program? Why does OSHA require its use?
11. What are safety data sheets? Why are they used on construction sites?
12. What is a phased project-specific accident prevention plan? Why is it prepared?
13. What are three objectives of a jobsite accident investigation?
14. What type information must be reported to OSHA regarding a construction injury in which a worker broke his or her leg?
15. What is EMR? How does it affect a contractor's labor costs?

## Exercises

1.  Prepare a job hazard analysis for the construction of the roofing on the NanoEngineering Building project.
2.  Develop a substance abuse policy letter for the NanoEngineering Building project.
3.  Develop the section of the phased accident prevention plan for the NanoEngineering Building project that covers the early site preparation phase of the project.
4.  What can you tell about a general contractor who has an EMR of 1.8? 0.4?
5.  Which construction trade is likely to have a higher worker's compensation base rate:
    a.  steel erector or landscaper?
    b.  drywall installer or roofer?
    c.  carpet installer or high voltage electrician?
6.  List five examples of passive safety control and five examples of active safety control.

# 15 | Contract change orders

## Introduction

A change order is an agreement signed by the general contractor, the architect, and the owner after the contract has been awarded, modifying some aspect of the scope of work and/or adjusting the contract sum or contract time, or both. Change orders also are known as *contract modifications*. They may be additive, if they add scope of work, or deductive, if they delete work items. Since they occur on most construction contracts, managing change orders is an important project management function. In this chapter, we will discuss the major issues involved in managing contract change orders. Change orders originate from a variety of sources and for a variety of reasons.

Box 15.1 lists some of the more common sources of change orders. The largest count of change orders usually comes from design errors, but they are not necessarily due to poor design. Sometimes designers do not have sufficient funds, resources, or time to do a complete design. Sometimes owners contract with the sub-consultants (such as mechanical or electrical engineers) directly and not through the architect. This may result in multi-discipline documents that are not fully coordinated. And design errors can occur due to the complexity of the project or innocent human error or oversight. Some design error change orders can be expensive, but often individually they are not costly. Conversely, the most expensive change orders come from owner-directive scope changes such as adding space or changing program or use of a building.

A third category of change orders stem from uncovering unknown site conditions, which can be the most difficult type of change order for an owner to accept. These generally result from inadequate site exploration prior to starting design. Hidden or latent conflicts or conditions are common in site work (for example, unknown buried debris) or remodeling (for example, rotten wood structure). When the contractor encounters hidden site conditions that adversely impact construction, the project manager must "promptly" notify the architect and owner and provide the architect an opportunity to make an inspection. Article 3.18.2 of the ConsensusDocs 500 contract contains this requirement.

---

## Box 15.1   Change order sources

- Architect generated scope changes, change in material specifications
- Contractor generated material or detail change or value engineering proposal
- Design errors, discrepancies, or lack of coordinated multi-discipline design
- Discovery of unknown or hidden conditions
- Municipality generated changes, code interpretations or permit review changes
- Not-in-contract equipment discrepancies, utility connections, size variations
- Owner scope changes, additions, or changes to the program
- Specified product is unavailable or no longer manufactured
- Interference from third party or another contractor not under the control of the general contractor
- Unusually adverse weather

---

The lack of coordination of not-in-contract, or owner-furnished, equipment and materials is one type of change order that is avoidable. Owners sometimes believe that they can contract directly with subcontractors and suppliers and save the general contractor's fee. This is common with items such as kitchen equipment, auditorium equipment, furnishings, custom casework, and landscaping. Problems arise, and the project is disrupted due to the lack of coordination of owner-furnished materials. Owners often will eventually pay much more to resolve these conflicts than they would have paid the general contractors in fees to manage this work.

## Change order proposals

Change order proposals (COPs) may be initiated either by the owner or by the general contractor. An owner-initiated COP request provides a description of the proposed change in the scope of work and requests that the contractor provide an estimate of additional cost (if the proposal adds scope of work) or deductive cost (if the proposal deletes scope of work). If the proposed change extends the project duration, the contractor also requests additional time when responding to the owner's COP request. A contractor-initiated COP generally results either from a differing site condition or from a response to a request for information (RFI). The contractor's proposal describes the proposed change to the scope of work and provides any requested adjustments to contract price and time. An example of a contractor's COP is shown in Box 15.2. The procedures used on the NanoEngineering Building are shown in Article 9 of the CM-at-risk contract included on the companion website. COPs do not modify the construction contract. Only change orders and construction change directives, which are discussed below, modify the contract.

Many COPs are initiated by subcontractors. Let's look at an example. The carpet subcontractor submits an RFI to the project manager stating that carpet is not clearly identified for room 291. The floor plan (drawing A2.04-TI) indicates 291 is a conference room, and the specifications indicate that all conference rooms receive loop carpet, but the finish schedule (drawing A10.01-TI) shows the floor finish in Room 291 is to be unfinished concrete. The carpet subcontractor

# Box 15.2    Change order proposal

## NORTHWEST CONSTRUCTION COMPANY
1242 First Avenue, Cascade, Washington 98202
(206) 239-1422
NanoEngineering Building

Change Order Proposal Number:  23                    Date:  10/27/16

Description of Work:        **Enclose pipes and conduits in pipe chase in the first floor lobby with hard-rated lid.**

Referenced Documents:     RFI #241, CD #7, and Sketch #16

| Estimate Summary: | | | Total |
|---|---|---|---|
| 1 | Direct Labor:  (See attached detail pricing recap) | | $ 1,000 |
| 2 | Supervision: | 15" of DL | $   150 |
| 3 | Labor Burden: | 50" of DL | $   500 |
| 4 | Safety: | 2" of DL | $    20 |
| 5 | Total Labor: | | $ 1,670 |
| | | | |
| 6 | Direct Materials and Equipment: (See attached pricing recap) | | $ 2,315 |
| 7 | Small Tools | 3" of DL | $    30 |
| 8 | Consumables | 3" of DL | $    30 |
| 9 | Total Materials and Equipment: | | $ 2,375 |
| 10 | Total Direct Work, Items 1-9 | | $ 4,045 |
| | | | |
| 11 | Subcontractors: (See attached subcontractor quotes) | | $15,000 |
| | | | |
| 12 | Overhead on Direct Work Items: | 0" | $     0 |
| 13 | Fee on Direct Work Items: | 3" | $   121 |
| 14 | Fee on Subcontractors: | 3" | $   450 |
| 15 | Subtotal Overhead and Fee: | | $   571 |
| 16 | Subtotal all Costs: | | $20,188 |
| | | | |
| 17 | Insurance and Excise Taxes | 2" | $   404 |
| 18 | **Total this COP:** | | $20,592 |

This COP may or may not have a schedule impact, the extent of which cannot be analyzed until the work is completed.

Approved by: *Jeffrey Jackson*                              11/5/16
            Jeffrey Jackson, University of Washington              Date

assumed that there was not any carpet in this room and has asked a question for verification. Another way this could have surfaced would have been through a submittal drawing of planned floor covering materials. Remember that well-written RFIs and thorough and timely submittals are methods for surfacing discrepancies and, therefore, are part of the active quality management process discussed in Chapter 13. The project manager should research the issue and forward the question or submittal to the architect. The architect responds to the question and indicates that the subcontractor's interpretation is incorrect and that carpet is required in Room 291.

The architect's response to the RFI may contain a request for a COP. If not, the subcontractor will notify the project manager regarding any perceived cost and time impacts from the architect's response. If the project manager believes a change order is warranted, he or she will send a change order request to the subcontractor using a form similar to the one shown in Chapter 6. The project manager uses the subcontractor's response to the change order request to prepare a change order proposal similar to the one illustrated previously in Box 15.2. The following guidelines should be used when preparing a change order proposal:

- each proposal should be numbered sequentially;
- the description of the proposed changes should be clear;
- direct labor, hours, labor burden, supervision costs are separated, itemized, and indicated;
- direct material and equipment costs should be summarized;
- subcontract costs are listed separately from direct costs;
- markups should include overhead (home office versus field), fees, taxes, insurance, contingency, and bonds; and
- a line should be provided for the owner to sign approval.

As with RFIs, submittals, and pay requests, the project manager should do everything possible to sell the change order proposal to the architect and owner. All relevant supporting documents should be attached. Some examples of attached documents include:

- the originating RFI or submittal,
- copies of relevant drawings and specifications and photographs,
- subcontractor and supplier quotations,
- all detailed quantity take-off and pricing recap sheets, and
- any relevant letters, memos, meeting notes, daily diaries, or phone records.

The project manager should require similar documentation from the subcontractors. All COPs should be tracked with a log. As soon as a potential changed condition arises, it should be assigned a number and entered in the log. An example COP log is shown in Worksheet 15.1.

The owner may accept and sign the proposal, which makes the rest of the process straightforward. Often, however, there will be questions related to the request and subsequent negotiations. This is to be expected and is part of working as a team. Remember that the proposal is usually negotiable. The project manager is not taking a hard-line approach at this time. A weekly change order proposal meeting outside the regular owner architect contractor construction coordination meeting is a good way to discuss and resolve change issues. It is beneficial to keep change order discussions out of the coordination meeting as the discussion of extra costs has a way of undermining the communication process.

# Worksheet 15.1 COP log

## NORTHWEST CONSTRUCTION COMPANY
1242 First Avenue, Cascade, Washington 98202
(206) 239-1422

## CHANGE ORDER PROPOSAL LOG

Project No.: 9821    Project Name: NanoEngineering Building    Project Manager: Ted Jones

| COP No. | Originating Document | Description | Originating Date | COP Date | Amount Requested | Date Approved | Approved Amount | CO No. | Comments |
|---|---|---|---|---|---|---|---|---|---|
| 1 | | Permit documents | 8/15/15 | 9/1/15 | 0 | 9/1/15 | 0 | 1 | No impact |
| 2 | IDC #1 | Over excavation for footings | 9/15/15 | 10/1/15 | $21,500 | 10/10/15 | $20,000 | 1 | |
| 3 | RFI, ASI | Rated lid in basement | 10/17/15 | 11/5/15 | $4,351 | 11/7/15 | $4,351 | 2 | |
| 4 | | Column rebar change | 10/12/15 | 10/15/15 | $222 | 11/15/15 | $222 | 2 | |
| 5 | RFI | Carpet manufacturer change | 10/12/15 | 11/1/15 | ($1,200) | 11/1/15 | ($1,200) | 2 | |
| 6 | Submittal | Toilet accessory backing | 11/1/15 | 11/15/15 | $475 | NA | NA | NA | Rejected |
| 7 | | Beam and duct conflict | 11/1/15 | | | | | | |
| 8 | CD #3 | Low voltage light controls | 11/15/15 | 12/1/15 | $3,500 | 12/1/15 | $3,600 | 2 | |
| 9 | CD #4, ASI | TI Package for basement | 11/16/15 | 11/16/15 | $155,550 | 12/15/15 | $139,007 | 2 | Negotiated |
| 22 | | Sprinkler and duct conflict | 10/1/16 | 10/1/16 | $11,575 | NA | NA | NA | Rejected |
| 23 | RFI 241, CD 7 | Ceiling in lobby | 10/22/16 | 10/27/16 | $20,592 | | | | Pending |

## Impact analysis

When analyzing the impact of a proposed change order, the project manager determines the direct cost of additional materials, labor, equipment, and subcontractors and the impact of the change on the scheduled completion of the project. If the change delays critical activities, additional contract time generally is justified. Time impacts are determined by updating the construction schedule by adding the additional scope of work contained in the proposed change order and evaluating its impact on the scheduled completion date for the project. If additional contract time is not granted, the project manager may need to accelerate some activities to compensate for the additional work. The jobsite indirect, or overhead, costs associated with these impacts also need to be estimated and included in the change order proposal price.

The timing and volume of change orders may disrupt the planned flow of work among project team members. This impact is difficult to quantify and may be equally difficult to negotiate with the owner. The best approach is to compare the as-built schedule with the as-planned schedule at the end of the project to determine what inefficiencies were created by the timing and volume of change orders. If adverse impacts occurred due to the timing and volume of change orders, the project manager may consider submitting a claim for compensation using the procedures described in the contract general conditions and discussed in Chapter 16.

## Pricing change orders

The ultimate goal for the project manager with respect to change orders is to have them approved so that all of the contractors can get paid. The easiest way to achieve this goal is to be realistic with respect to pricing on direct and subcontract cost estimates. Overly inflated prices will only delay the process. Quantity measurements generally are verifiable and should not be inflated. Wage rates paid to craft employees are verifiable and should not be inflated. Subcontractor quotes should be passed through as is without adjustment (unless incomplete) from the project manager to the owner. The subcontractors and suppliers should practice the same procedures with their second- and third-tier firms. Labor productivity rates should be derived from preapproved resources. Material prices should be actual and verified with invoices or quotations. Any deviations in the foregoing suggestions may damage team-build relationships and trust among the parties.

Mark-ups, or percentage add-ons, are used to recover indirect costs such as:

- bonds,
- cleanup,
- consumables,
- detailing,
- dumpsters or rubbish removal,
- fee on direct work,
- fee on subcontracted work,
- field overhead,
- foremen costs,
- hoisting,
- home office overhead,
- insurance,

- labor burden,
- material transportation and handling,
- safety equipment,
- small tools,
- supervision costs,
- taxes, and
- contingency.

It is possible to see a series of mark-ups that could essentially double the hard cost of the direct work. Owners often get frustrated with these add-ons. They do not understand why they have to pay more for the change than simply the direct costs. Many of these items are required, and sometimes it is just the presentation that makes them difficult to sell. There are several ways to smooth the process and reinforce fairness for all parties.

Sometimes, general contractors (and subcontractors) are asked to propose mark-up rates with their bid proposals. This eliminates the need to negotiate a mark-up rate for change orders once they occur. Some mark-ups are stated in the contract documents. This is another reason to carefully read the contract prior to preparing the estimate. Some designers will try to lump the subcontractor and general contractor mark-ups together in the supplemental conditions to prevent mark-ups on top of mark-ups. This is difficult for the GC to manage and not necessarily fair to all of the contractors.

Another system is for the project manager to negotiate mark-up rates with each subcontractor and the owner soon after receiving the notice to proceed but before many changes have occurred. This will avoid negotiating mark-up rates for each change order separately.

Mark-up rates for general contractors' fees typically range from 3 percent (larger commercial GCs) to 10 percent (residential or tenant improvement projects). This rate is usually the same percentage fee that was used on the original estimate and is stated in the contract. Home office overhead costs are assumed to be included in this fee. Jobsite general conditions are not usually allowed unless they can be substantiated on individual change issues or it can be proven that the scope of the change will extend the project schedule. In order for the project manager to prove that additional general conditions and/or time are warranted, he or she needs to provide documentation to back up the claim, such as a very detailed construction schedule. Contractor attempts of recovery for additional jobsite general conditions, time extensions, home office general conditions, and impact costs are often contentious, especially in the COP process, and more often fall into the claim category as discussed in Chapter 16.

Subcontractors tend to receive higher fees because their volumes of work generally are less than those of general contractors and their labor risk, which is determined by the ratio of direct labor to subcontract value, is higher. Subcontractors may receive a 10 percent fee and an additional 10 percent overhead. Both of these rates will depend upon how many of the items listed earlier in this section are anticipated to be included in the base estimate, or are in the fee or overhead, or are allowed in addition to the fee.

## Change order process

The two documents used to modify construction contracts are the change order and the construction change directive. Both documents generally are prepared by the architect. If the architect or owner and the project manager have negotiated a mutually acceptable adjustment in the contract price, time, or both, a change order is executed modifying both the scope of work and the terms of the contract. If a mutually acceptable adjustment has not been negotiated for an owner-initiated change, the owner may choose to withdraw the change order proposal request. Section 00-72-00.7 of the supplemental specifications contains the change order procedures for the NanoEngineering Building. Contract Article 9 contains additional instructions regarding making changes to the contract. We have included an example developed by the owner of this case study project as Box 15.3. Formal copyrighted ConsensusDocs Form 202 and AIA document G701 are common formal change order coversheets and preferred by many contractors, clients, and architects.

A formal contract change order may be used to incorporate several approved change order proposals into the construction contract. Some owners choose to issue monthly formal change orders incorporating all change order proposals approved during the month. COPs do not modify the terms of the contract, and are not added to the schedule of values for pay purposes until incorporated into the contract by formal contract change order. The contract documents should be annotated with all scope changes contained in each change order. Once the change order has been signed by the architect, owner, and contractor, the contract scope, price, and time have been modified. The contractors' signature indicates that they agree that the adjustment in price and time adequately compensate for the added scope of work. After the general contractor's contract has been modified, the project manager should modify appropriate subcontracts and purchase orders for major material suppliers.

The change order is used when the owner and the project manager have agreed to an appropriate adjustment to the contract price and time. If the owner does not agree to a change order proposal submitted by the project manager, the project manager may withdraw the proposal or may submit a claim using procedures that will be discussed in Chapter 16. If the owner and the project manager have not negotiated a mutually agreeable contract adjustment but the owner still wants to proceed with the change, an AIA construction change directive (CCD), Form G714, often is issued. A CCD is used when the owner and the project manager have not agreed on an appropriate adjustment to the contract price and time. The CCD may also be used when the change must be implemented before the project manager has had time to evaluate its cost and time impacts. An alternative to the AIA CCD is an interim directed change (IDC) ConsensusDocs Form 203. Box 15.4 is a customized client-originated "change directive" version of the CCD and IDC for our case study project.

The CCD modifies the scope of the construction contract and may adjust the contract price. Often construction change directives indicate that the general contractor is to proceed with the changes and that an appropriate adjustment to contract price and time will be negotiated later. These are sometimes referred to as *un-priced change orders*. Upon receipt of a construction change directive, the project manager determines the impact of the change on the project and submits a change order proposal requesting an appropriate adjustment of contract price and time. Also upon receipt of a construction change directive, the project manager must issue appropriate direction to affected subcontractors and suppliers. The CCD changes the contract scope when signed by the architect and owner, but it does not change the contract price or time unless also signed by the contractor. If the contractor refuses to sign the construction change directive, any change in contract price and time will occur only after the architect, owner, and contractor have negotiated a separate change order.

## Box 15.3    Contract change order

### CONTRACT CHANGE ORDER
Where the basis of payment is a Guaranteed Maximum Price

**Client:**      *University of Washington*
**Address:**   *1101 15<sup>th</sup> Avenue, Seattle, WA 98011, (206) 239-3556*

**Contractor:**   *Northwest Construction Company*
**Address:**      *1242 First Avenue, Seattle, WA 98202, (206) 239-1422*

**Project:**   *NanoEngineering Building*          **Project Number:**   *9821*

**Change Order Number:** *2*          **Change Order Date:**   *12/21/15*

| | |
|---|---|
| Original Contract Price: | $61,024,000 |
| Previously Approved Change Orders:    Through Change #: *1* | $     20,000 |
| Previous Contract Price: | $61,044,000 |

Incorporate the following agreed Change Order Proposals. See attached COP log.

| 3 | Rated Lid | $   4,351 | |
|---|---|---|---|
| 4 | Column Rebar | $     222 | |
| 5 | Carpet | $ (1,200) | |
| 8 | Low Voltage Lighting | $   3,600 | |
| 9 | T.I. Finishes | $139,007 | |
| Total added/reduced/unchanged this CO: $145,980 | | | $    145,980 |
| Current and revised Contract Price: | | | $61,189,980 |

Previous Contracted Schedule Completion:
     1/15/2017
Net change per this CO:    *0*     Days added/reduced/unchanged:          0
Current and Revised Contracted Schedule Completion:          1/15/2017

**Contractor's Approval:** *Ted Jones*          Date: *12/21/15*
**Architect's Approval:** *Norm Riley*          Date: *12/21/15*
**Client's Approval:** *Jeffrey Jackson*          Date: *12/21/15*

## Box 15.4   Change directive

### CHANGE DIRECTIVE
Where the basis of payment is a Guaranteed Maximum Price

**Client:** *University of Washington*
**Address:** *1101 15ᵗʰ Avenue, Seattle, WA 98011, (206) 239-3556*

**Contractor:** *Northwest Construction Company*
**Address:** *1242 First Avenue, Seattle, WA 98202, (206) 239-1422*

**Project:** *NanoEngineering Building*          **Project Number:** *9821*

**Change Directive Number:** *7*          **Change Directive Date:** *10/22/16*

In conformance with ConsensusDocs® Contract 500, Article 9.2 and in conjunction with special conditions Article 00-50-00, the Client provides the following direction to the contractor in regards to the following changed conditions:

*According to ZGF's response to RFI 241, the contractor is hereby directed to incorporate the added ceiling in the elevator lobby. This change was originated by the Fire Marshal.*

The client directs the contractor to incorporate this Change Directive into the work as follows:

X     Develop a COP cost estimate to be negotiated. Do not yet proceed with the work
___   Proceed with the work on a T&M basis. Keep track of all costs
___   Proceed with the work, cost and time impacts to be negotiated later
___   Proceed with the work. There is not cost or time impact associated with this CD
___   Proceed with the work according to the architect's estimate for $_____

**Contractor's Approval:** *Ted Jones*          Date: *10/22/16*
**Architect's Approval:** *Norm Riley*          Date: *10/22/16*
**Client's Approval:** *Jeffrey Jackson*          Date: *10/22/16*

## Contract issues

Before embarking on the bid process or the proposal process, it is imperative that the general contractor's project manager and officer-in-charge carefully review the contract included in the request for proposal or request for quotations along with all the front-end or division 00 special conditions of the specifications. Some of the issues that may arise relating to change orders include the following.

- Notice: Contracts require a strict time-frame from when a contractor realizes a potential change of scope has occurred and when he or she must (1) notify the architect or client of the change and (2) must have it priced and presented to the architect or client. Terms such as "As Soon As Possible" or "Immediately" or "Promptly" are unadvisable and subject to interpretation. Time frames such as within 7 or 14 or 21 days are easier to enforce. ConsensusDocs 500 Article 9.4 requires 14 days.
- Differing site conditions: Contractors are required to discover conditions either on the jobsite or between the separate contract documents that are potentially different from what was represented. If a contractor proceeds with work that is in conflict with the architect's and client's intention, then the contractor is taking the risk of an incorrect interpretation and may be required to perform rework at no increase in time or cost. As discussed prior, RFIs are the tool most commonly utilized to provide notice. Our case study contract Article 3.3.5 covers this.
- Response to RFIs: Some RFI formats dictated by the client or architect may require the contractor to present cost and time impacts at the time the RFI is asked. This is difficult for contractors to do until they have received a response. In addition, once the RFI has been answered, the architect and client may assume there are not any cost and schedule impacts unless the contractor responds within a stipulated time frame, similar to Notice above.

Ultimately, the contractor must understand these clauses and be willing to accept the risk, if any, or include appropriate additional funds in their estimate or proposal in the form of contingencies or fees. If contractors feel the risks are too great, they can choose to pass on this perspective project.

## Summary

Change orders occur on most construction projects. The most common causes are design errors, owner-directed scope changes, and differing site conditions. COPs may be initiated either by the owner or by the project manager. An owner-initiated proposal request contains a description of the proposed change in the scope of work and requests that the project manager provide an estimate of the cost and time impact on the project. A project manager–initiated proposal describes the proposed change in the scope of work and requests an appropriate adjustment to the contract price and or time. The project manager maintains a change order proposal log to track all change order proposals whether generated by the owner or by the project manager. Once the project manager and the owner have negotiated a mutually agreeable adjustment in contract price and time, a formal change order is executed, modifying the contract. If the owner wants to proceed with a change but has not negotiated an appropriate adjustment to the contract price and time, a CCD is issued, directing the work to be accomplished. The project manager later submits a change order proposal responding to the CCD with a proposed adjustment to the contract price and time.

## Review questions

1.  What is a change order to a construction contract?
2.  What are five causes of change orders on a typical construction project?
3.  Why does using owner-furnished equipment on a project sometimes result in change orders?
4.  What is the difference between an owner-initiated change order proposal and a contractor-initiated change order proposal?
5.  How do designer responses to field questions sometimes result in change orders?
6.  What factors should a project manager consider when analyzing the impact of a proposed change order on a construction project?
7.  What recourse does a project manager have if the owner denies a change order proposal?
8.  How does a change order proposal become a change order?
9.  What is the difference between a change order and a construction change directive?
10. What action does a project manager take upon receipt of a construction change directive?
11. Why might an owner decide to issue a construction change directive?

## Exercises

1.  Prepare a COP with a value greater than $10,000. Include a cover letter that allows the owner to sign and approve. Include all relevant supporting documents such as quantity sheets, recap sheets, subcontractor quotes, sketches, RFIs, submittals, and memoranda.
2.  Prepare a contract change order incorporating the change order proposal prepared in exercise 1. Note the schedule impact, if any.

# 16
# Claims and disputes

## Introduction

Sometimes issues occur on construction projects that cannot be resolved among project participants. Such issues from a contractor's perspective typically involve requests for additional money or time for work performed beyond that required by the construction contract. The project manager first submits a change order proposal for the contract adjustment. Once the owner and the project manager agree, a change order is negotiated and executed, adjusting the contract duration and/or price. A claim will, therefore, not be filed. The procedures for developing and processing change order proposals (COPs) and change orders were discussed in Chapter 15. If the owner does not agree with the project manager's request, the result may become a contract claim. Procedures for processing claims typically are prescribed in the contract. Articles 9.4 and 13 of the ConsensusDocs 500 contract on the companion website outline the procedures used for a CM-at-risk scenario, such as was the case with the NanoEngineering Building. The normal procedure is for the project manager to formally submit the request for additional compensation and/or time to the owner or the architect along with documentation supporting the project manager's position. The owner or the architect formally responds to the contractor's request agreeing, agreeing in part, or rejecting the contractor's claim. If the project manager does not agree with the response, the claim becomes a contract dispute, and one of the dispute resolution techniques discussed later in this chapter is used to settle it. It is to the advantage of both the contractor and the owner to resolve the dispute quickly. The farther up the ladder above the project level the dispute reaches, the more likely it is to become adversarial, time-consuming, and expensive. The partnering and team-building techniques discussed in Chapter 5 and others have been adopted by many in the construction industry in an attempt to reduce the number of claims and disputes. The frequent meetings and the issue escalation system are used to resolve issues at the project level in a timely manner.

## Sources of claims

Construction claims typically result from one of the following causes:

- constructive acceleration, or compression, where the contractor is required to perform a task or tasks at a faster rate or with additional manpower than planned;
- constructive changes, where the scope of work was modified by the owner;
- cumulative impact of numerous RFIs and change orders;
- defective or deficient contract documents;
- delay caused by the owner or the owner's agents;
- site conditions that differ from those described in the contract documents;
- unresponsive contract administration; and
- weather impacts.

Any of these issues may affect the project manager's and superintendent's ability to complete the project within budget and the prescribed duration. The major issues in claims and disputes are:

- What are the issues?
- Who is responsible?
- What does the contract dictate?
- What are the cost and time impacts?

Time impacts are determined by assessing the impact of the disputed work on the construction schedule. Cost impacts are determined by evaluating the actual costs for any additional work and the overhead costs incurred by the contractor for being on the project any extra time.

## Legal impacts

As indicated, claims differ from change orders in that negotiated and executed change orders resolve and incorporate into the contract new or changed conditions. Individual claims often involve more money and time impacts than do change order proposals. Some of the industry trends and concepts that impact the size and resolution of claims follow.

- *Waiving rights to claim*: There can be clauses in the contract, lien releases, pay requests, and executed change orders where general contractors, and in-turn subcontractors, are required to waive their rights to future claims. Different jurisdictions and courts may either uphold these clauses or deem them unlawful.
- *No claim for delay*: This is a contract clause that allows the owner or designer to delay a project due to either additional scope or delayed decisions, but the contractors are not allowed to recover additional jobsite or home office overhead costs, even though they may be given time extensions on the contract schedule.
- *Eichleay formula*: Eichleay is a complicated and controversial element of some claims. This formula calculates the home office overhead and company fee potential that are applicable to an individual project. If that project is delayed at no fault of the contractor, then in addition

to recovering jobsite general conditions costs, home office costs are added to the claim as well.

- *Case law*: As discussed with litigation below, construction disputes are civil, not criminal, law suits. Attorneys and the judge will rely on prior court cases to base their position, and ultimately decision, on this case: If a court found in favor of the contractor on unknown site conditions on one project, that finding may be applied to this project as well.
- *Notice requirements*: Notice was discussed in Chapter 15. It is applicable to claims as well. If an owner denies a contractor's change order proposal, then, as dictated by contract, the contractor has a set amount of time, say 24 days, in which to notify the owner that he or she disagrees with the disposition of the COP and intends to file a claim. The claim itself will also need to be submitted formally to the owner by the given time frame and/or referred to mediation or arbitration as discussed in the next section of this chapter. Essentially the contractor cannot wait until the end of the project, or later, to bring up old issues.

Following are additional terms and concepts that affect both development and resolution of many construction claims.

- *Quantum versus merit*: The merit or validity of a claim should be analyzed before the quantum or amount or cost of the claim.
- *Excusable versus non-excusable*: If the contractor caused the problem, he or she is deemed non-excusable and cannot claim for additional time and money.
- *Compensable versus non-compensable*: A compensable claim is one by which the contractor is deemed eligible for financial remedy.
- *Culpable or culpability*: This determines the degree to which one of the parties may be partially or fully responsible for delay or cost impact.

## Dispute resolution techniques

There are several methods used in the construction industry to resolve disputes. Prevention or avoidance is the least costly and most efficient and should, of course, be at the top of the list. If a dispute arises, negotiation should be the first technique attempted because it is relatively inexpensive and not time-consuming. If negotiation does not result in resolution, other techniques may be selected in an attempt to find resolution. Litigation or going to court usually is the last resort, as it is the most expensive and time-consuming. Mediation, arbitration, and dispute resolution boards (DRBs) are techniques that have been adopted by the construction industry as alternatives to litigation. These three are considered alternative dispute resolution (ADR) techniques. Box 16.1 ranks the various resolution techniques in terms of cost and time, with negotiation being the least costly and litigation the most costly.

## Negotiation

Negotiation involves both parties to the dispute sitting down, discussing both sides of the issue, and reaching agreement on an appropriate resolution. This is the most efficient and least

---

### Box 16.1   Dispute resolution techniques

**Dispute Resolution Techniques
Ranked in Order of Increasing Cost
and Time to Resolution**

1. Negotiation

2. Mediation

3. Dispute Resolution Board

4. Arbitration

5. Litigation

---

expensive method of resolution and does not rely on any outside support. Negotiation should be attempted as soon as the dispute surfaces to avoid creating an adversarial relationship between the contractor's field personnel and those of the owner. Early negotiation may prove successful if attempted before the parties have had a chance to formulate strong positions regarding the dispute or involving legal advisors.

## Mediation

Mediation is an assisted negotiation process in which settlement discussions are facilitated by a neutral third party. Both the contractor and the owner may agree to bring in an outside mediator, or it may be required by the construction contract. Article 13.4 of the ConsensusDocs 500 contract requires mediation prior to taking a dispute to arbitration. The individual selected as mediator must be acceptable to both parties. As a consequence, this person must be credible and knowledgeable of the issues in question. The mediator listens to both parties' positions and attempts to help them reach a consensus regarding resolution. The mediator does not render any decision on the issues but serves as an intermediary between the parties to the dispute. Most mediation sessions last only 1 day, but occasionally they may extend into a second day.

Prior to the mediation session, both parties submit brief memorandums to the mediator setting forth their positions with regard to the issues that need to be resolved. Mediation starts with an initial joint session during which each side presents its position regarding the dispute. After the opening statements are completed, the representatives of the contractor and those of the owner are assigned to different rooms. Once assigned to different rooms, the parties are not assembled together again until a resolution has been reached. The mediator meets separately with each party and discusses the strengths and weaknesses of each position in an attempt to craft a mutual agreement. Once both parties to the dispute reach agreement, which is usually based upon compromise, the mediator prepares a formal agreement document that the parties sign.

For mediation to be successful, both parties must have a desire to reach a settlement, the representatives participating must have authority to settle, and the mediator must have the trust of both parties. Not all mediations are successful; sometimes the mediator is unable to identify an acceptable solution. Mediation sessions are kept confidential, and there is no written record kept of the process.

## Arbitration

Arbitration is a more formalized alternative than mediation but still more informal than litigation. In this case, the disputing parties present their case to a neutral third party, called the *arbitrator*, who is empowered to make a decision. Arbitration can be binding or nonbinding. Binding arbitration means that the arbitrator's decision is final, and nonbinding arbitration means that either party may pursue litigation after arbitration. If the losing party in nonbinding arbitration decides to litigate, the arbitration award may be entered as evidence in court. Both the contractor and the owner may agree to take their dispute to arbitration, or it may be required by the construction contract. Article 13.5 of our example contract requires the use of binding arbitration to resolve disputes. Arbitration is considered a more efficient process than litigation because it is quicker and less expensive. Even so, arbitration of large, complicated cases can still be time-consuming and expensive. In addition, arbitrators may have more technical expertise in the subject matter under dispute than do judges, but not always.

Most arbitration of construction disputes is conducted in accordance with the Construction Industry Arbitration Rules of the American Arbitration Association (AAA). Arbitration is initiated by the claimant (for example, the contractor) giving written notice to the respondent (for example, the owner) and the AAA regional office of its intent to demand arbitration of a dispute. The notice contains a description of the nature of the dispute and the remedy requested. Upon receipt of the written notice, the AAA regional office furnishes both parties a list of qualified, potential arbitrators with biographical information regarding each candidate. If both the claimant and the respondent agree to an arbitrator, the AAA will appoint the selected individual. Often, one party selects three candidates from the list, and the other party selects one from the list of three. If the parties cannot agree on an arbitrator, the AAA will designate one. Most members of the AAA are licensed attorneys. A single arbitrator is used in most instances, but a panel of three arbitrators may be used on large, complex disputes. When the panel is used, one member is selected by the contractor, and a second is selected by the owner. The third member is selected by the other two members, similar to DRBs below. See the AAA for additional details at *www.adr.org*.

Arbitration hearings are similar in many respects to a trial. Both parties make opening statements and present their case to the arbitrator or arbitrators. Case presentations may include witnesses, expert reports, depositions, and documentation for evidence. Rules of evidence and procedure are less formal in arbitration than they are in litigation. The arbitrator is the judge of the relevance and admissibility of the evidence offered. Written transcripts of the proceedings generally are not required. Either mediation or arbitration before litigation is often required by the contract and may also be required by the courts, essentially directing the parties to attempt to resolve their dispute themselves before taking the court's time. Another form of ADR, very similar to arbitration, is a mini-trial, which is not as widely utilized today in favor of the others mentioned above and below.

## Dispute resolution board

A DRB, or the Board, is quite different from the other techniques used for dispute resolution in that the Board is organized before construction starts and any disputes arise. This proactive approach provides a panel of experienced construction professionals who are knowledgeable of the specific project. A DRB was first utilized in the mid-1960s on the Boundary Dam project near Spokane, Washington. The board consists of three industry experts selected for their knowledge of the type of construction being accomplished. One member is nominated by the project owner, a second by the contractor, and the third by the other two members. All board members must be acceptable to both the owner and the contractor. The nomination and approval process follows.

1.  The general contractor nominates board member 1 and submits to the client for their approval.
2.  The client nominates board member 2 and submits to the general contractor for its approval.
3.  Board members 1 and 2, once approved by both parties, nominate board member 3 and submit to both the general contractor and the client for their mutual approval.
4.  Board member 3, after having been mutually approved, usually becomes the chair of the dispute resolution board.

The decision to use a DRB generally is made by the owner. If one is to be used, the construction contract may include a lengthy specification in the supplemental conditions, as was the situation with our case study project. Box 16.2 is an example of a DRB specification clause. ConsensusDocs 200.4 and 200.5 are other choices. Once the board members are selected, they sign a three-party agreement obligating them to serve both parties fairly and equally. The agreement is signed by each board member, the contractor, and the owner; it is a three-party agreement, but with five signatures.

The board members are provided copies of the contract plans and specifications to become familiar with the project and make periodic site visits to keep abreast of construction progress. Typically board members meet on the project site monthly or quarterly to review construction progress and hear any issues in dispute. Either the contractor or the owner may refer an issue to the DRB for a recommendation. The board review consists of an informal hearing at which each party explains its position and an examination of all appropriate documentation and evidence. After deliberation, the board renders a written nonbinding recommendation for resolving the dispute together with an explanation of how the board arrived at its conclusions. It is up to the owner and the contractor to accept or reject the board's recommendation. If the recommendation is not accepted, the dispute would be resolved by mediation, arbitration, or litigation as prescribed by the contract. The DRB recommendations are usually available for subsequent proceedings, but if the claim reaches all the way to litigation, more than 95 percent of the courts find in concert with the previous DRB recommendation.

DRBs are more prevalent on public than private projects. Although there are cases where DRBs have been used successfully on private projects, many private clients prefer to negotiate and avoid

# Box 16.2    DRB specification

## DISPUTE RESOLUTION BOARD SPECIFICATION

1. A dispute resolution board (hereinafter 'Board') will be established to assist in the resolution of disputes in connection with, or arising out of, performance of the work of this contract.
2. Either the owner or the contractor may refer a dispute to the Board. Such referral shall be initiated as soon as it appears that the normal dispute resolution procedure is not working and prior to initiation other dispute resolution techniques.
3. The Board shall consist of one member selected by the owner and one member selected by the contractor, with these two members to select the third member who will serve as the Board Chair. All three members shall be mutually acceptable to both the owner and the contractor. No member shall have a financial interest in the work, except for payment for services as a Board member. The Board will be considered established once the attached Three-Party Agreement among the owner, the contractor, and all three Board members has been executed. This agreement establishes the scope of Board services and the rights and responsibilities of the parities.
4. The Board will formulate its own rules of operation, except to the extent any such rules are provided herein or in the Agreement. To keep abreast of the work, the members shall regularly visit the project, keep a current file, and regularly meet with the other members of the Board and representatives of the owner and contractor. The frequency of these visits shall be as established in the Agreement.
5. Compensation for Board members and Board operating expenses shall be shared by the owner and the contractor equally. Based upon the expected schedule and scope of the _NanoEngineering Building Project_ the owner and the contractor shall include in their respective budgets a stipulated line item totaling _$50,000 each_. If for some reason the scope of the Board increased, or the process extends beyond the anticipated schedule of _22_ months, then the parties agree to mutually update this budget.
6. Requests for Board review of a dispute shall be submitted in writing to the Board Char and the other party. The request for review shall state in full detail the specific issues of the dispute and provide full documentation. The Board Chair will request a response from the other party and establish a hearing schedule that provides adequate time for both parties, as well as the Board members, to review the documentation. During the hearing both parties will have an opportunity to present their position and respond to the other party's position. The Board may or may not ask questions of the parties and may or may not request additional supporting documentation. After the hearing the Board will meet separately and provide a written response and recommendation to the parties to resolve the dispute.
7. Board recommendations are not binding on either the contractor or the owner. Within two weeks if either party disagrees with the Board's recommendation, that party will be required to provide written notice to the other party and the Board and pursue other alternative dispute resolutions as prescribed in the contract. The Board's finding and recommendation will be made available for review in subsequent proceedings.

formal dispute resolution. Research into private DRBs was the focus of this author's master's thesis, *The Applicability of Dispute Resolution Boards on Privately Financed Construction Projects*.

Many DRB project board members are also members of the international Disputes Resolution Board Foundation (DRBF). The DRBF has developed guidelines for establishing and operating boards and also conducts training seminars not only for potential board members but for owners, architects, and contractors. Additional information about the DRBF can be found at their website, *www.drb.org*.

Very similar to a DRB, the parties may appoint a single third-party "neutral" who may serve the entire duration of the project as the three-member board does or comes in just in the case of a dispute, similar to mediation. Our case study contract article 13.3 offered the parties a choice of a single third-party neutral or a DRB.

## Litigation

Litigation means referring the disputed issue to a court for resolution. This involves hiring legal counsel, preparing necessary documentation, and scheduling an appearance before a judge. This typically is the most time-consuming and expensive, as previously shown in Box 16.1. Most contractors and owners attempt to resolve their disputes without resorting to litigation, but it might be used as a last resort. Litigation is not an expedient means of dispute resolution. It can take years before the matter proceeds to trial and, if appeals are made, the final result will be delayed further.

Litigation starts when the plaintiff (for example, the contractor) submits a formal complaint to the appropriate court. The complaint contains the names and addresses of the parties, the allegations that form the basis of the complaint, and the nature of the relief sought. The defendant (for example, the project owner) then submits a formal denial of the allegations contained in the plaintiff's complaint.

Before the case goes to trial, both parties go through a discovery process. Discovery refers to a number of procedures available to both parties prior to trial, the purpose of which is to learn as much as possible about the case. Procedures available include interrogatories, discovery of documents, expert witness reports, deposition of witnesses, and inspection of the project site. Interrogatories are a series of written questions sent by one party to the other. Written responses must be provided within a specified time. Discovery of documents means that each party has a right to request copies of all pertinent files and documents of the other side. A deposition is a cross-examination of a witness taken under oath prior to trial. A verbatim record of the questions and answers is made by a court reporter.

Once discovery is completed, the complaint is scheduled for trial. A construction lawsuit is a civil proceeding rather than a criminal proceeding. The trial may be conducted by a judge without a jury or, on demand of either party, the trial may be held before a jury. The purpose of the trial is first to determine the facts in the case and then to apply case law to render a decision. Court trials are formal proceedings, and every word spoken is recorded verbatim by a court reporter. The judge controls the type of exhibits and testimony that go into the trial record as evidence. There are strict rules regarding what is admissible and what is not. Each side is allowed to present its case starting with the plaintiff, followed by the defendant, who presents a rebuttal. Once all the evidence has been presented, either the jury deliberates and renders a decision or the proceedings are closed and the judge prepares a written decision. The judge's decision may not be issued until months after the trial. Either party has a right to appeal the trial court's decision to an appellate court that will consequently increase costs and delay final resolution.

## Summary

Sometimes issues occur on construction projects that cannot be resolved between the contractor and the owner. Such issues result in claims that are formally submitted by the contractor using the procedures specified in the construction contract. Unresolved claims that become disputes should be resolved in the most efficient, least costly manner. Several techniques for dispute resolution have been adopted by the construction industry.

Negotiation involves both parties' sitting down, discussing the issue, and reaching an appropriate resolution. Mediation is an assisted negotiation process in which both parties agree to use a neutral facilitator, or mediator. The mediator listens to both parties' positions and attempts to help them reach a consensus regarding resolution.

Arbitration involves both parties' presenting their case to a neutral third party for a decision. The arbitrator requests written position papers and supporting documentation from both parties to the dispute. Once the documentation has been reviewed, the arbitrator conducts a hearing allowing both sides to present their positions and renders a written decision.

A dispute resolution board is a panel of three industry experts convened prior to the start of construction. The board periodically meets at the project site to review progress on the project, and hear any issues in dispute. Either the contractor or the owner can refer an issue to the dispute resolution board for a recommendation. Informal, but comprehensive, hearings are held, and the board makes a recommendation. Board recommendations are not binding on the contractor or owner. A single third-party neutral is similar to a DRB but an even more economical option for the contracting parties.

Litigation means referring the dispute to a court for resolution. This involves hiring legal counsel, preparing necessary documentation, and scheduling an appearance in court. This typically is extremely time-consuming and expensive. Most contractors and owners attempt to resolve their disputes without resorting to litigation, but it might be used as a last resort.

## Review questions

1. What are three potential causes of claims on a construction project?
2. What is mediation, and how is it conducted?
3. What are the differences between mediation and litigation? What are the similarities?
4. What is arbitration, and how is it conducted?
5. What is a dispute resolution board? When are its members selected?
6. Why do dispute resolution board members make periodic visits to the construction project site?
7. What would be the advantages and disadvantages of a single third-party neutral to (a) a DRB or (b) mediation?
8. Why do courts typically find in favor of previous DRB recommendations?
9. What is the discovery process that is used in litigation?
10. Why is litigation generally the most expensive of the methods of dispute resolution discussed in this chapter?

## Exercises

1. Assume a contractor has a contract for the construction of an elementary school. Assume our sample contract applies here. The contractor's project manager believes the contractor is owed additional time and money because of an owner-caused delay. What procedure should the project manager use in submitting the claim?

2. How can the partnering techniques discussed in Chapter 5 be used to resolve issues between the contractor and the owner?

3. Assume you are the project manager for Northwest Construction Company on the NanoEngineering Building and have a dispute with the owner. You and the owner have decided to use mediation to attempt to resolve the dispute. What experience and abilities are you looking for when you select the mediator?

4. As project manager for Northwest Construction Company for the NanoEngineering Building, you are required to nominate a member for the dispute resolution board. What criteria would you use to select an individual to nominate?

5. As project manager for Northwest Construction Company for the NanoEngineering Building, you have been asked to prepare an agenda for the quarterly site meeting of the dispute resolution board. What agenda would you propose?

6. Assume you are the project manager for a construction company with an average annual volume of $500 million, which spends 1 percent on home office overhead and routinely realizes 2 percent in pre-tax pure profits. Your own project is scheduled to last 2 years and has a contract value of $100 million, 5 percent of which is for jobsite general conditions. Midway during construction, your client has delayed your project for 1 year at no fault of your own. Prepare two claims:

    a. Jobsite overhead costs. Can you recover 100 percent? What would you do with your people and equipment during this 1-year shut-down?

    b. Utilize the Eichleay formula to prepare a claim for impacts to home office overhead and lost fee potential.

# 17 Project close-out

## Introduction

When physical construction on a project nears completion, the project manager (PM) develops a project close-out plan to manage the numerous activities involved in closing out the project. Just as the project start-up activities described in Chapter 8 are essential when initiating work on a project, good project close-out procedures are essential to timely completion of all contractual requirements and receipt of final payment. Not only is efficient project close-out good for the general contractor, it is good for the owner. The project manager wants to close out a project quickly and move on to another project. Minimizing the duration of close-out activities generally maximizes the profit on a project, as it limits project overhead costs and facilitates early receipt of final payment and any retention. Efficient close-out and turnover procedures also minimize the contractor's interference with the owner's move-in and start-up activities. Project managers sometimes lose credibility with owners because of inefficient close-out procedures. Dissatisfied owners rarely give contractors a second chance but award future projects to other contractors.

Project close-out is the process of completing all the construction tasks and all documentation required to close out the contract and consider the job complete. Article 10.6 of our case study contract addresses requirements for substantial completion and final payment. Close-out unfortunately can take up to a year after the receipt of a certificate of occupancy. For purposes of discussion, we have divided the close-out process into the following six categories:

- commissioning,
- construction close-out,
- contract close-out,
- financial close-out,
- project manager's close-out, and
- warranty management.

## Commissioning

Commissioning is a series of interrelated actions taken to ensure that a constructed project meets the owner's requirements. These actions typically involve a range of services to minimize surprises when the owner assumes control and operates the completed facility. A commissioning plan is developed during the design phase that outlines the commissioning process to be used. Primary commissioning activities include verification testing of selected equipment and systems, training of the owner's operating and maintenance personnel, and identification of any defects or warranty issues that may be realized in the first year or two of operation. It does not take the place of normal start-up and testing and balancing of the mechanical, electrical, and plumbing (MEP) systems. Equipment and systems verification tests generally are conducted using checklists developed for the specific project based on the owner's requirements.

During commissioning, the MEP systems undergo aggressive testing that is intended to replicate all four seasons the building will experience. "Design days" or extreme warm or cold or humid days are simulated. The commissioning process may be performed by an outside separate third-party commissioning agent who is contracted direct to the owner, but many mechanical subcontractors also sell themselves to owners as qualified to do this work. Some owners who have experienced in-house facilities personnel may occasionally take the point on commissioning as well. This author participated on one of the first third-party commissioned biotechnology research facility projects. Regardless of who takes the lead, the commissioning team should include:

- owner's representative,
- owner's facilities personnel,
- mechanical and electrical engineers,
- general contractor's PM or project engineer,
- MEP subcontractor PMs and/or superintendents,
- third-party commissioning agent, and
- MEP equipment representatives.

A separate specification section will define the roles and responsibilities of the commissioning team. The cost of an independent third-party commissioning agent may be in the range of 1 percent of the total construction cost. In the case of the NanoEngineering Building, the contract required a "Commissioning Authority" who did not provide any labor, material, or equipment to assist with the commission process but rather acted more as an observer or advisor. The contractor and its subcontractors provided all the commissioning effort as part of their individual scopes of work; therefore, the total cost of the commissioning process on that project would be difficult to determine, but the general contractor's contribution was included in Worksheet 3.5. Special conditions specification sections 01-75-00 and 01-91-00 defined those roles and processes on that project.

## Construction close-out

Many of the items related to completing the construction on a project and closing out the physical aspects are the responsibility of the project superintendent, similar to the physical mobilization of the project discussed in Chapter 8. Major construction activities associated with physical close-out include:

- ensuring that all construction tasks have been completed in accordance with contract requirements and noting deficiencies on the punch list;
- obtaining certificates of substantial completion and occupancy; and
- demobilizing.

## Punch list

Throughout the construction of the project, the superintendent and quality control staff should be inspecting completed work and identifying any deficiencies as part of the active quality management program discussed in Chapter 13. When the project is nearing completion, the PM requests a pre-final inspection from the owner and architect. The list of deficiencies identified during this inspection is known as the *punch list*. An example punch list is shown in Box 17.1.

---

### Box 17.1   Punch list

#### PUNCH LIST

| | | | |
|---|---|---|---|
| Project: | *Middle School* | Inspection Date: | *12/5/16* |
| Area/System: | *Exterior/Shell/Site* | Status Date: | *12/10/16* |

Inspection Participants:   *Nancy Rogers, Jason Brown, Jim Johnson, Maggie Prince*

| Item | Description | Complete? | Verified? |
|---|---|---|---|
| 1 | Downspout splash blocks not installed | NIC | OK |
| 2 | Front door hardware requires adjustment | Yes | OK |
| 3 | Phase II site work pending spring weather | | |
| 4 | Parking lot requires sweeping | Yes | OK |
| 5 | Two handicap signs to be squared and straightened | Yes | |
| 6 | Re-paint entrance arrows, too light | Yes | OK |
| 7 | Fuel oil tank certificate requires exterior display | Yes | |
| 8 | Remove temp silt fence along stream (next spring) | | |
| 9 | Remove GC's project sign | Yes | OK |
| 10 | Add building address for Fire Marshal | Yes | |
| 11 | Remove dumpster | | |
| 12 | Clean exterior of windows | Yes | OK |
| 13 | Incorrect door mat installed: Check specifications | | |
| 14 | General site clean-up | Ongoing | Partial |

---

All items on the punch list must be corrected before the project can be considered complete. The punch list usually is developed by the architect, but some owners prepare their own. Some consultant design team members, such as the mechanical or electrical engineers, may also develop punch lists. Many separate groups within an owner's organization may develop their own separate punch lists depending upon their specialization. How many punch lists should a PM receive? He or she should try to receive, and manage, just one. If numerous lists are developed, items often will be overlooked, duplicated, or lost or even contradict one another. The best method is for all parties who are interested in inspecting the project to gather and walk the jobsite together. One punch list form should be used. The architect, or designee, can collect all of the input and issue the list to the project manager. The PM or superintendent may also develop the punch list with input from all the interested parties.

The inspection should occur early enough to allow the superintendent and subcontractors sufficient time to complete the corrections prior to the owner taking occupancy. If the team is still developing the punch list after the owner has moved in, it is difficult to determine who did what damage. It also is a mistake to start the process too early. If there still are basic construction activities to complete, there may be additional damage that was not listed on the original punch list. The punch list should be prepared as soon as all major construction activities are completed.

The general contractor and subcontractors should take no more than 2 weeks to complete all of the work on the punch list. If the process takes too long, the responsible parties may have demobilized, and it may be difficult to get them back to the jobsite. The punch list should be signed off by the responsible parties as deficiencies are corrected. A copy of the annotated punch list should be sent to the architect. The architect may then wish to revisit each punch list item and verify its completion or may perform spot checks on items or rooms. The architect will then notify the owner that all deficiencies on the punch list have been corrected.

Sometimes items will appear on the punch list that are outside the scope of the contract, or at least the PM may believe so. The superintendent should not perform punch list work that is not required by the contract and then later request a change order to pay for it. The PM should respond to the architect and owner as soon as possible so that they can decide whether they want to pay to have the extra work performed. There ultimately may be disagreements regarding some of these items. The PM will need to meet with the owner and work out any differences prior to the project's being completely closed out.

Early punch lists or pre-punch lists were discussed in Chapter 13 as a means of active quality control. Additionally, some contractors may choose top-down finishing as a method of physically closing the project down, minimizing rework, reducing the size of the punch list, and ultimately saving costs.

## Certificates of completion

There are two basic certificates of completion. The *certificate of substantial completion* is issued by the architect. It indicates that the project is sufficiently completed such that it can be used for its intended purpose but that there still may be some minor deficiencies that need correction. Substantial completion is addressed in Article 10.6 of our sample contract. All items on the punch list may not have been corrected, but the architect agrees that the facility can be used. A copy of the corrected or updated punch list should be attached to this certificate. An example

ConsensusDocs 814 certificate of substantial completion is included on the companion website. Substantial completion is a significant contractual event. It ends the general contractor's liability for liquidated damages. The contractual completion date stated in the contract agreement is the date by which substantial completion is to be achieved. Article 6.6 of the contract agreement on the companion website states this requirement for the NanoEngineering Building.

The *certificate of occupancy* (C of O) is issued by the city, county, or municipality having jurisdiction over the project site. This is usually the same agency that issued the original building permit. It signifies that all code-related issues have been accounted for and that the building is approved and safe to occupy. It is often a formality that follows completion of all the various inspections and subsequent approval of all other construction permits. The most important aspect of the C of O is inspection by the fire marshal for life-safety issues. The project may have minor deficiencies that are not related to life-safety and still be approved for occupancy. For example, the landscaping may not yet be installed because it is the middle of December and there are three feet of snow on the ground. In this case, the PM may have to post a bond with the city which guarantees completion of this portion of the project even though the C of O was approved. Some public agencies will issue a temporary certificate of occupancy (TCO) that allows the owner conditional use of the building, for a stated period of time, say 6 months, while other non-critical work such as landscaping or some interior finishes may be completed at a later time. An example C of O for the NanoEngineering Building is shown in Box 17.2. Receipt of either just the certificate of substantial completion or the C of O is not enough to signify the project is complete; the PM needs to obtain both.

## Occupancy

When will the owner occupy the completed project? The owner will usually occupy his or her new building as soon as the city allows it to be occupied, whether it is totally finished or is not. Joint occupancy occurs when the owner accepts and occupies a portion of the building while the contractor is still working in another portion. This may be dictated by the owner's need to begin business in the new facility by a certain date, or the owner may want to begin moving equipment into the new facility. Joint, dual, or conditional occupancies often are unavoidable and may be undesirable if not managed properly. The PM and superintendent may have problems differentiating the owner's cleanup and routine maintenance from construction punch list and warranty work. Payment of permanent utility bills also may be an issue. Labor problems may occur when construction and operations personnel begin to work together in a building, especially if one party is union and the other is not. It is better for all parties if occupancy is delayed until the certificates of completion have been obtained and all items on the punch list have been corrected. If the contractor and the owner are to have joint occupancy, the PM should work with the owner to establish good procedures early and to communicate clearly about each other's expectations. The owner's right to take partial occupancy of an uncompleted project is described in Article 10.7 of our case study contract that states in part:

> *The owner may occupy...when work is designated in a certificate of substantial completion...and public authorities authorize occupancy.*

## Box 17.2    Certificate of occupancy

# *Certificate of Occupancy*

### City of Seattle, Washington

### *NanoEngineering Building*

This Certificate issued pursuant to the requirement of Section 306 of the International Building Code, as amended, certifying that at the time of issuance this structure was in compliance with various ordinances of the City of Seattle, regulating building construction or use for the following:

| | |
|---|---|
| Use Classification: | *Educational Building, Classrooms, Laboratories* |
| Permit No. | *BLD97-00514* |
| Occupancy Group: | *B, A3, S2, Low Hazard Storage* |
| Construction Type: | *1A, Fully Sprinkled* |
| Owner of Building: | *University of Washington* |
| Owner Address: | *4100 15$^{th}$ Ave NE, Seattle, WA, 98102* |
| Building Address: | *1701 NE Grant LN* |
| Legal Description: | *The West ½ of the NW ¼ and the NW 1/4 of the SW ¼ lying East of 15$^{th}$ Ave NE* |
| Building Official: | *Christian Roust* |
| Date Issued: | *January 7, 2017* |

## Demobilization

Although demobilizing, or physically moving off the site, sounds easy, it can involve considerable work and be expensive, especially to the general contractor. At that time, there may not be funds that can be dedicated to a foreman and a small crew to clean up the site. Similar to mobilization, most estimators do not put a line item in their original estimate for demobilization. The question always arises as to whose garbage is piled up. Seldom do subcontractors claim the surplus cardboard or pallets. Similar to the discussion regarding joint occupancy, garbage accumulation can become mixed between the construction crews and the owner, as the owner is busy moving in. Demobilization involves closing the project office and removing all contractor-owned materials and equipment from the project site. Utility companies are notified to disconnect temporary utilities, and vendors are notified to close project accounts. The project staff are phased out and reassigned to other projects. The project files are collected and taken to the construction firm's records holding area.

## Contract close-out

One of the first steps in contract close-out is to check the specifications for close-out requirements. Ideally this happens during project start-up so that requirements are delineated in each subcontract and major purchase order. Section 01-77-00 of the specifications contain the close-out requirements for the NanoEngineering Building. The PM should make a list of what he or she thinks needs to be done and then ask the designer for verification, similar to an early submittal log. Some of the major items the project manager will be required to prepare and submit as part of the contract close-out process are discussed in this section.

# As-built drawings

As-built drawings are one of the most significant close-out deliverables. They were introduced in Chapter 9. Actual dimensions and conditions of the installed work are noted on the contract drawings. Requests for information and sketches attached to the back of the preceding drawing, as discussed in Chapter 10, can be a big help. The PM should collect all the as-built drawings from the subcontractors. Mechanical, electrical, and civil disciplines are of the utmost importance because they include hidden systems that would cause severe damage if cut or may need to be accessed during a future remodel. Usually the general contractor will develop as-built drawings for the architectural, structural, and (sometimes) the civil portions of the work. The best person to mark up the as-built drawings is the foreman or assistant superintendent who oversaw the work. The as-built drawings should be submitted to the owner using the same procedures as were discussed in previous chapters for shop drawings. Many projects that utilized building information models or computer-aided design in the design process may also require the contractors to record as-built conditions electronically. In this case, an additional specialized individual is often required on the contractor's team and must be considered when developing the jobsite general conditions estimate as discussed in Chapter 3.

# Operation and maintenance manuals

Operation and maintenance (O & M) manuals are the collection and organization of manufacturers' data regarding operating procedures and preventive maintenance and repair procedures for all the operational equipment as well as many of the finish items. This information should be collected from all the subcontractors and suppliers and organized in binders to be presented to the owner. Sometimes the contract specifications dictate the organization of these documents. Section 01-77-00, subsection 1.3.B of the specifications describes the required organization of the O & M manuals for the NanoEngineeing Building. It states in part:

*The O&M Manuals shall contain all the information needed to operate, maintain and repair all systems, equipment, and product finishes provided in the Project. They shall be presented and arranged logically for efficient use by Owner's operation personnel.*

The O & M manuals should be processed as a submittal using the procedures discussed in Chapter 7 to request designer approval. This is sometimes required and always a good idea. Some designers will request that draft copies be submitted for comment prior to submission of the final copy. This also should be well received by the construction team. The PM cannot allow the project team simply to collect previous submittal data and place them into a three-ring binder. Similar to as-built drawings discussed above, the O & M manuals may be required to be submitted electronically. The information needed by the owner is not sales propaganda but O & M tools. Project managers need to ensure that the O & M manuals fully conform to contract specifications. Contracts often require that O & M manuals contain detailed descriptions of how each building system operates, which is more than a collection of manufacturers' literature describing system components.

## Warranties

The PM collects written, original, and signed warranties and guarantees from all the subcontractors and suppliers. An example subcontractor warranty is included later in this chapter as Box 17.3. The owner may also request an overall warranty document from the general contractor. The warranties can all be inserted as a section in the O & M manuals.

## Test reports

Throughout the course of the job, various materials, installations, and systems were tested. This includes balancing reports for the mechanical systems. All these test reports, along with commissioning documentation, also may be bound in the O & M manuals. Accurate record keeping and diligent filing throughout the course of the project will help with this.

## Sustainability

The contract for our case study project stated that the contractor is to "assemble and maintain records to document Leadership in Energy and Environmental Design (LEED) goal compliance." Assembly of LEED certification documentation, or any other sustainability standard, is the responsibility of the general contractor's PM as introduced in Chapter 5. However, as stated, many of these close-out activities are delegated to a project engineer or, in this case, a LEED-certified project engineer or LEED coordinator.

## Extra materials

The specifications may require that extra quantities, say 1 percent to 3 percent, of some finish materials be supplied to the owner upon completion. This would include materials such as paint (of each color used), ceramic tile, carpet, and ceiling tile. This assists the owner with future repairs and remodels. Occasionally, contractors need to be reminded that these surplus materials are not to be used for punch list or change order work.

# Permits

The PM should require each subcontractor to provide an approved and signed permit from the authority having jurisdiction over that trade. This indicates that the subcontractor's work was performed in conformance with code requirements. The PM also may be required by contract to submit to the designer all interim inspection reports or signature cards received from the city or county throughout the course of the job. Signed-off permits also may be included as a section in the O & Ms.

# Close-out log

How does the PM track completion of the O & M manual from the plumbing subcontractor, the warranty from the electrical subcontractor, the surplus rubber base material, the touch-up paint included on the painting subcontractor's punchlist, and the lien release from the drywall subcontractor? It is done by use of a close-out log similar to the one shown in Worksheet 17.1. This log should be developed early in the project. The PM should review the specification sections for each of the subcontractors and list all close-out items required. The log should be issued to the subcontractors and suppliers early and often to remind them of their contractual responsibilities. A general letter can be sent out, repeatedly if necessary, listing all open issues. Many subcontractors are slow at closing out their portions of the project. Remember our earlier discussions on timely pay request submissions. The PM must work closely with the subcontractors and suppliers to get the project closed out so that all parties may receive their retention.

Some of the reasons the PM should pursue a timely close-out are as follows:

- Officially end the clock on potential liquidated damages.
- Identify late subcontractor change orders and close the door for future claims.
- Begin the clock on warranties and guarantees.
- Maintain good relations with the owner, designer, and subcontractors. The PM will work with all these people again on another project. It is important that everyone leave the project with a sense of teamwork and accomplishment. The sooner the client can achieve the C of O, the quicker they can transfer from a construction loan to a permanent loan and save significant interest costs.
- Minimize project overhead costs. Using the PM and superintendent to oversee project close-out can be expensive, and some contractors rely on project engineers and foremen to manage close-out. This is not necessarily a good practice and may alienate owners.
- Receive the final progress payment (different than retention). Although the last progress payment should not be tied to close-out, some owners may hold it as an incentive for timely close-out. Remember: Cash is a useful tool. The PM should help the owner by being proactive in the close-out process.
- Close out the subcontracts and purchase orders as soon as possible. It is not recommended that the PM hold their retention any longer than the owner holds the general contractor's retention. This is to ensure fair treatment of these key members of the construction team and close out the subcontractors before they suffer financial hardship due to insufficient cash flow.

# Worksheet 17.1   Close-out log

## SUBCONTRACTOR AND SUPPLIER CLOSE-OUT LOG

Project No.: 9821    Project Name: NanoEngineering Building    Project Manager: Ted Jones

| Description | Site Work | Supply Doors | Utilities | Plumbing | HVAC | Electrical | Drywall | Tile | Flooring | Paint | Insulation | Supply Rebar |
|---|---|---|---|---|---|---|---|---|---|---|---|---|
| Complete punch list | 10/1/16 | NA | 10/5/16 | | | | | | | | | NA |
| Sign final change order | 10/1/16 | 11/13/16 | 10/10/16 | | | | | | | | | 12/1/16 |
| Turn in permits | NA | NA | NA | | | | NA | NA | NA | NA | NA | NA |
| Extra materials | NA | NA | NA | NA | NA | | NA | | | | NA | NA |
| O&M manuals | NA | | | | | | NA | | | NA | NA | NA |
| As-built drawings | NA | | | | | | | | | | | |
| Final lien release | 10/15/16 | | | | | | | | | | | |
| Second tier lien releases | 10/15/16 | | | | | | | | | | | |
| Test certificates | NA | | | | | | | | | | | |
| Union affidavits | 10/15/16 | | | | | | | | | | | |
| Back charges clear | Yes | | | | | | | | | | | |
| Demobilized | Yes | | | | | | | | | | | |
| Warranties | NA | | | | | | | | | | | |
| Retention released | 11/1/16 | | | | | | | | | | | |

- Receive release of retention. The retention held may be approximately equal to the fee on the project. Therefore, the PM cannot realize a profit on the project until retention has been collected. The request for release of retention is prepared and submitted using the same procedures as those discussed in Chapter 11 for monthly payment requests. Sometimes, this is best resolved by keeping the final pay request, which bills up to 100 percent complete, separate from the release of retention request.

## Financial close-out

The financial close-out of a project is not the accountant's or the officer-in-charge's responsibility, although they often become involved, nor is it the project engineer's responsibility. The PM is responsible for financially closing out the project with both the owner and his or her subcontractors. The items involved in financial close-out of a project are discussed in this section.

## Change orders

All change order proposals and construction change directives must be negotiated and settled prior to financially closing out the contract. After the PM receives the final change order from the owner, he or she should issue a final modification to each subcontractor and supplier. The word *final* is important. Even though a subcontractor may not have any cost issues in the last change order with the owner, the PM should still issue a final modification to each subcontract and major purchase order for $0.00. In this way they are officially notified that there will not be any other opportunities to collect on change orders. The final subcontract change order should be issued on a form similar to Box 6.7 with the addition of the word *final* in several locations.

## Application for final payment

The request for final payment is submitted at final completion once all items on the punch list have been completed and all required documentation has been submitted. The request is for the release of all retention as well as for any work completed since the preceding payment request was submitted. A final lien release and a waiver of claims generally are required to accompany the application for final payment.

## Lien releases

The general contractor will be required to provide a final and unconditional lien release to the owner, and most likely the owner will require similar releases from all subcontractors. Regardless of whether the owner requires them, the PM should obtain final and unconditional lien releases from all subcontractors and suppliers before submitting the final lien release to the owner. Unconditional lien releases were discussed in Chapter 11 as a part of the request for payment process. The final lien release ends the contractor's right to place a lien on the project. It indicates that all financial accounts with the owner have been settled.

All materialman's notices should be reviewed, and each subcontractor and supplier should be required to obtain final and unconditional lien releases from all their second- and third-tier subcontractors and suppliers, regardless of their contract values. If a company was concerned enough to reserve its lien rights, then it needs to release those rights in order to receive its retention. The contractor's release, without final releases from the subcontractors and suppliers, is without much value. Even though the contractor has released its rights, the subcontractors can still place a lien on the job for alleged lack of payment. It is because of this that most experienced owners also will require the PM to collect as many final releases from subcontractors and suppliers as possible and forward them to the owner before release of retention.

## Project manager's close-out

There are a few additional close-out activities that are the responsibility of the project manager.

- Preparing an as-built estimate: If it is not maintained throughout construction or prepared near the completion of the project, it never will be developed. Considerable work went into tracking actual costs. This is valuable input to the firm's ongoing ability to improve its estimating accuracy. Input of the as-built estimate into the firm's estimating database is necessary, if the database is to be kept current. Many project managers will simply input actual costs alongside the original quantities to the company's database, but the most accurate historical data takes into consideration actual hours and actual material costs and actual quantities, such as:

    $25,000 of redi-mix concrete cost @ 200 CY = $125/CY purchase price

    10,000 hours for D/F/HW labor @ 900 doors = 11 MH/opening

- Preparing an as-built schedule from the data marked up on the jobsite meeting room schedule: as discussed in Chapter 4.
- Preparing an analysis of the final forecast of project costs versus contract amount and the resultant fee: A narrative should be developed with a chart tracking the fee forecast throughout the project. This lessons-learned paper will help the PM on subsequent similar projects, maybe with the same owner or designer.
- Conducting a post-project meeting with the owner and designers: They should be asked to fill out a report card that serves as an evaluation of the construction firm's performance.
- Conducting a post-project meeting with all the field and office project personnel and the officer-in charge: The meeting should include any staff personnel who may have some input, for example safety, accounting, procurement, estimating, and scheduling. Lessons learned should be applied on future projects.

## Warranty management

Once the owner has accepted the completed project, he or she wants assurance that the project was completed in accordance with contract requirements. This assurance generally is provided in the form of a warranty. Article 3.10 of our example contract discusses warranty. It states in part that the contractor warrants or guarantees that:

> *All materials and equipment used on the project were new and met contract requirements, and...all work is free from defects and meets contract requirements.*

The warranty period required by most construction contracts is 1 year from substantial completion. Warranty would start earlier on any portions of the project occupied by the owner prior to substantial completion. Longer warranties may be required in the technical specifications for selected components, such as roofing, glazing, elevators, or electric motors. General contractors typically impose the same warranty requirements on their subcontractors and suppliers as are contained in the general contract with the owner.

As discussed previously, the collection of subcontractor and supplier warranties is a part of project close-out activities. All warranties should be submitted by subcontractors and suppliers on their own letterhead using text similar to the example shown in Box 17.3. Warranty submission should be monitored and tracked in the close-out log as illustrated in Worksheet 17.1. Final payments to subcontractors and suppliers should not be made until warranty documents have been received. Supplier warranties may be inserted in the O & M materials provided to the owner. Subcontractor performance bonds should not be cancelled until the end of the warranty period to protect against subcontractor failure to respond to warranty claims.

## Warranty service requests

The PM needs to organize a responsive warranty service that:

*   satisfies the customer,
*   ensures warranty costs are reasonable,
*   minimizes conflicts, and
*   maximizes the potential for repeat business.

When a warranty claim is made, the PM should review the project records and determine which subcontractors need to be notified. Appropriate subcontractors should immediately be notified orally and in writing that the owner has made a warranty claim. If the claim involves contractor-performed work, appropriate craftspeople should be sent to the site to inspect and make acceptable repairs. In emergency situations, the PM may tell the owner to notify the appropriate subcontractor directly and then notify the project manager. The PM needs to be aware of all warranty claims so that he or she can ensure they have been resolved. Warranty procedures need to be explained by the PM to both the owner and the subcontractors during contract close-out so that all parties understand the procedures to be used.

The PM needs a system for tracking all warranty claims. A log similar to the one illustrated in Worksheet 17.2 can be used. Once the work on a warranty claim has been completed, the owner should be asked to sign a release indicating a satisfactory remedy. This tells the PM that the warranty claim has been resolved.

## Box 17.3   Warranty

### *Richardson Electric, Inc.*
*800 Pacific Highway*
*Seattle, Washington 98902*

WARRANTY FOR:  Electrical Work

We hereby warrant and guarantee all electrical work which we have installed and/or furnished in the NanoEngineering Building project in Seattle, Washington is in compliance with the contract documents for one year from the date of substantial completion.

We agree to repair or replace to the satisfaction of the Architect all work that may prove defective in workmanship or materials within that period, ordinary wear and tear and unusual abuse or neglect excepted, together with all other work which may have been damaged or displaced in so doing.

All repairs or replacements shall have a warranty period equal to the original warranty period as herein stated, dated from the final acceptance or repairs or replacement.

In the event of our failure to comply with the above-mentioned conditions within a reasonable time after being notified in writing, we collectively and separately do hereby authorize the Owner to proceed to have defects repaired and made good at our expense, and will pay the costs and charges therefore immediately upon demand.

Date:  January 11, 2017

*Rick Richardson*
Richardson Electric, Inc.

*Ted Jones*
Northwest Construction Company

## Warranty response as customer service

Just as customers expect quality work on their projects, they also expect timely response to warranty calls. Warranty response is an important aspect of customer service. Contractors who are poor at responding to warranty calls may not be invited to submit proposals on future projects. Subcontractors who are not responsive to warranty calls should not be used on future projects,

# Worksheet 17.2    Warranty log

## WARRANTY SERVICE LOG

Project Name: *NanoEngineering Building*                Project Manager: *Ted Jones*
Project No.: *9821*

| Item | Date Claim Received | Date Subcontractor Notified | Date Claim Resolved | Remarks |
|---|---|---|---|---|
| *Fan in Lab 280B inoperable* | *July 10, 2017* | *July 19, 2017* | *July 22, 2017* | *Fan replaced* |
| | | | | |
| | | | | |

because their poor performance may affect the customer's opinion of the general contractor's warranty support. A reputation for good warranty support can be an important tool when marketing the construction firm's services. The best way to minimize warranty claims and associated costs is an active quality management program as was discussed in Chapter 13. Using quality materials and skilled craftspeople who produce quality work should result in fewer warranty claims.

Project managers should contact owners just before the expiration of the warranty period and arrange for a joint inspection (owner and project manager) of the completed project. The purpose of the inspection is to identify any items that should be corrected under warranty. Owners tend to be impressed by the project manager's initiative and may be influenced to use the same construction firm on future projects.

## Summary

Project close-out is the process of completing all the construction tasks and all the documentation required to close out the contract and consider the project complete. The PM and the superintendent work closely together to ensure close-out procedures are comprehensive and efficient. Good close-out procedures typically result in higher contractor profits and satisfied owners. Issues relating to physical completion of construction are the responsibility of the superintendent, while management of close-out documentation is the responsibility of the project manager.

When the project is nearing completion, the PM asks the owner and architect to conduct a pre-final inspection. Any deficiencies noted during this inspection are placed on the punch list for future re-inspection. All deficiencies on the punch list must be corrected before the contract can be closed out. A significant project milestone is achieving substantial completion, which indicates that the project can be used for its intended purpose. The architect decides when the project is substantially complete and issues a certificate of substantial completion. Even though the project is substantially complete, the owner cannot move in until a C of O is issued by the local permitting authority.

The PM is responsible for the financial and contractual close-out of the project. This involves issuing final change orders to subcontractors and major suppliers and securing final and unconditional lien releases from them. As-built drawings, O & M manuals, warranties, and test reports must be submitted. The PM should develop a close-out log early in the project to manage the timely submission of all close-out documents. The objective is to close-out all project activities expeditiously so the contractor can receive the final payment and release of retention. As a part of close-out activities, the PM should conduct a post-project analysis to determine owner satisfaction and any lessons learned that can be applied on future projects.

Most construction contracts contain a warranty provision that requires the contractor to repair any defective work or replace any defective equipment identified by the owner within one year after substantial completion. Similar provisions should be included in all subcontracts to cover work performed by subcontractors. Subcontractor warranty documentation is collected during project close-out. Warranty claims need to be investigated and resolved. Generally, the owner submits any claims to the general contractor, who either notifies the appropriate subcontractors or sends an appropriate craftsperson to investigate. The contractor would send their own craftsperson only when the claim involves work that was performed by the contractor's own workforce. The PM should maintain a log to track all warranty claims and ensure their timely resolution. Warranty response is an important aspect of customer service. Poor support may jeopardize the contractor's ability to obtain future projects from the owner.

## Review questions

1. What is a punch list? When is it developed?
2. What action should the PM take if he or she believes that some of the punch list items are outside the scope of the contract?
3. What does substantial completion mean in the context of a construction contract?
4. Why is substantial completion a significant contractual milestone?
5. Who issues the certificate of substantial completion?
6. What is a certificate of occupancy?
7. Who issues a certificate of occupancy?
8. What are final and unconditional lien releases?
9. Why do some owners require that project managers submit final and unconditional lien releases from subcontractors?
10. Where does the PM obtain the information for the O & M manuals that are submitted to the designer for approval?
11. What items are tracked by the PM on a close-out log?
12. What type of post-project analyses are performed as part of the project manager's close-out activities on a project?
13. What does the warranty provision in a construction contract cover?
14. When does the warranty period begin?
15. From whom should the PM obtain written warranties?
16. Where would you determine which components require warranties that are longer than the 1 year required in the general conditions?
17. What is a warranty claim?
18. Why is warranty response an important customer service?

## Exercises

1.  Prepare a close-out log for a project. Include at least 10 subcontract and supplier categories.
2.  Prepare a final and unconditional lien release for Richardson Electrical for the NanoEngineering Building.
3.  Prepare a request for release for retention on the NanoEngineering Building.
4.  Assume you were the PM for the construction of a new high school. The school principal has made a warranty claim regarding the failure of a ventilating unit. You have notified the mechanical subcontractor, but the subcontractor has failed to take any action. What action should you take?
5.  Draw a flow chart of the material procurement process starting at the selection of the supplier and ending with receipt of the product warranty.

# 18

# Business aspects of project management

## Introduction

The performance of the project team in managing the completion of a construction project significantly influences the economic performance of the construction company. In this chapter, we will examine three basic business topics that are important to both the project team as well as to the successful operation of a construction company.

The first topic is business development. Construction companies often have a business development staff to identify prospective customers, but usually the project team needs to make the sale by convincing the prospective customer that the team will provide the best value in executing the contract and managing the project. Also, a construction company's reputation is greatly influenced by the performance of its project teams as the company aspires to do repeat business with desired customers.

The second topic is financial management. Two aspects are discussed: financial analysis of a construction firm and cash flow management. The project team's ability to manage cash flow will affect the overall financial condition of the construction company as will properly developing bids and proposals for prospective construction contracts.

The third topic is human resource management responsibilities of the project manager. The design of the project management organization was discussed in Chapter 1, and here we will discuss staffing, performance measurement, and employee development.

## Business development

Business development is the process of acquiring business for a construction company. It means retaining customers the company wishes to retain as well as acquiring new customers with whom the company desires to do business. The basic components of business development are marketing and sales. Marketing is the process of retaining desired existing customers, identifying potential

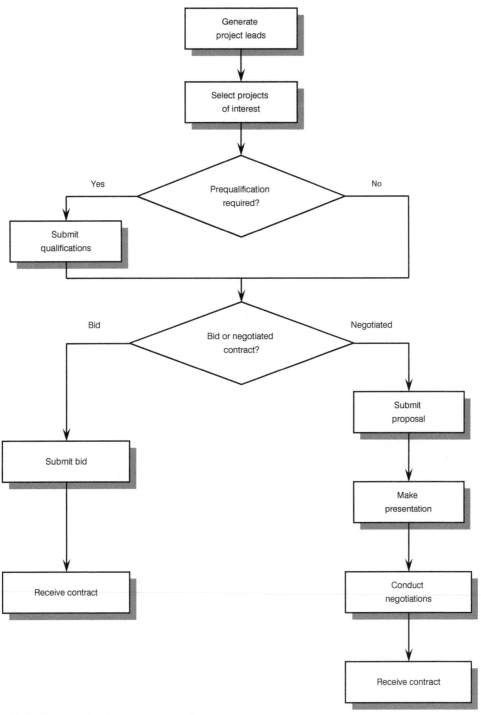

**Figure 18.1**   Business development process for construction services

new ones, and attracting new customers. Sales is contacting a specific potential customer and winning a construction contract.

The business development process for construction services is shown in Figure 18.1. The process parallels the bid and negotiated procurement methods discussed in Chapter 2. With the exception of some speculative residential builders who market products (homes), most construction companies market services, either construction services or design and construction services. To be successful, construction companies must understand the service characteristics desired by their customers and provide those service characteristics. These characteristics often relate to cost, quality, safety, service, and timeliness.

Most construction companies have business development staff members who gather information about prospective customers and projects and provide the information to company executives, project managers, and superintendents who actually make the sales. Most potential customers want to purchase services from people, not companies, so the project team is critical in obtaining a contract. Often, interviews with project teams are required as part of the prospective customer's selection process. To establish rapport with major customers, project managers often are assigned responsibility for cultivating and nurturing relationships with customers.

The most important marketing resources are the construction company's reputation and its satisfied customers, both of which are significantly affected by the company's project teams. Success depends upon understanding the process used by potential customers in selecting contractors for their projects and understanding the customers' specific project requirements. Retaining customers depends on the project team's ensuring customer satisfaction so that there may be opportunities for repeat business.

## Financial management

Good financial management skills are essential in the management of a construction company. The company must generate sufficient resources to be able to cover its financial obligations so as to remain in business. The project team's ability to manage cash on each project is a critical component of the company's financial management strategy.

Financial statements are used to provide a picture of the financial condition of a company. They are used as part of the subcontractor prequalification process discussed in Chapter 6 as well as by bankers, surety agents, and prospective customers in evaluating the financial condition of a construction company. The two most commonly used financial statements are the balance sheet and the income statement.

The balance sheet represents the financial condition of the company as of the date of the balance sheet, usually the end of a month, quarter, or year. An example balance sheet for a specialty contractor is shown in Box 18.1. Current assets are non-depreciable assets, such as cash, accounts receivable, and inventory. Fixed assets are depreciable assets that will be retained for longer than 1 year. Current liabilities are debts to be paid in 1 year. Accounts payable are debts owed to suppliers and subcontractors or for the purchase of any good or service.

The income statement summarizes the profitability of the company over a period of time, usually a year, showing all revenues and expenses. The following equations summarize the data contained in the income statement shown in Box 18.2

# Box 18.1   Example balance sheet

**Eastside Excavating Company**
**Balance Sheet**

December 31, 2016

ASSETS:

**Current Assets**

| | |
|---|---|
| Cash | $2,024,188 |
| Accounts Receivable – Trade | $4,657,875 |
| Accounts Receivable – Retention | $430,500 |
| Inventory | $89,463 |
| Prepaid Expenses | $51,923 |
| **Total Current Assets** | $7,253,948 |

**Fixed Assets**

| | |
|---|---|
| Land | $414,725 |
| Buildings | $1,964,925 |
| Furniture and Equipment | $377,468 |
| Motor Vehicles | $561,938 |
| Construction Equipment | $3,917,500 |
| Subtotal | $7,236,555 |
| Less Accumulated Depreciation | ($2,313,350) |
| **Total Fixed Assets** | $4,923,205 |

| | |
|---|---|
| **TOTAL ASSETS:** | $12,177,153 |

LIABILITIES AND OWNERS' EQUITY:

**Current Liabilities**

| | |
|---|---|
| Notes Payable | $250,000 |
| Accounts Payable – Trade | $2,417,625 |
| Accounts Payable – Retention | $89,000 |
| Accrued Payables | $285,875 |
| Accrued Taxes | $169,875 |
| Current Maturity – Long Term Debt | $312,500 |
| **Total Current Liabilities** | $3,524,875 |

| | |
|---|---|
| **Long Term Debt** | $1,125,000 |
| **TOTAL LIABILITIES:** | $4,649,875 |

**OWNERS' EQUITY**

| | |
|---|---|
| Capital Stock | $1,250,000 |
| Retained Earnings | $6,277,278 |
| **TOTAL OWNERS' EQUITY:** | $7,527,278 |
| **TOTAL LIABILITIES AND OWNERS' EQUITY:** | $12,177,153 |

## Box 18.2   Example income statement

**Eastside Excavating Company**
**Income Statement**

Calendar Year 2016

| | |
|---|---|
| **Revenues:** | |
| Sales | $22,698,572 |
| Cost of Sales | ($17,854,395) |
| Gross Profit | $4,854,177 |
| | |
| **Operating Expenses:** | |
| Financing | $250,875 |
| Marketing | $376,980 |
| General and Administrative | $2,796,845 |
| Operating Expenses | $3,424,700 |
| | |
| **Net Profit before Taxes:** | $1,419,477 |
| State and Federal Taxes | ($236,579) |
| | |
| **Net Profit after Taxes:** | $1,182,898 |

- Gross profit = sales – cost of sales
- Net profit before taxes = gross profit – operating expenses
- Net profit after taxes = net profit before taxes – taxes

Analysis of the *balance sheet* and *income statement* provides an assessment of the financial condition of the company. The primary financial characteristics of interest in the financial statements are liquidity, profitability, and leverage. Ratio analysis can be used to assess these characteristics at a single point in time.

Liquidity measures the ability of the company to meet its current obligations when they become due. The current ratio is used to measure the company's ability to pay current liabilities and have sufficient working capital to continue operations. It is defined as:

$$\text{Current ratio} = \frac{\text{Current assets}}{\text{Current liabilities}}$$

A typical current ratio for a commercial construction company should be at least 1.5. Companies with low current ratios may have difficulty in meeting their financial obligations.

Profitability is the ability of a company to generate profits from its operations. The return on assets can be used as a measure of the profitability of a company and is defined as:

$$\text{Return on assets} = \frac{\text{Net profit after taxes}}{\text{Total assets}}$$

A typical return on assets ratio for a commercial construction company should be at least 6 percent.

Leverage is the amount of debt being used to finance the company's operation. The debt-to-equity ratio is used as a measure of the leverage of a company and is defined as:

$$\text{Debt-to-equity ratio} = \frac{\text{Total liabilities}}{\text{Owner's equity}}$$

A typical debt-to-equity ratio for a commercial construction company should be less than 1.5. Companies with high ratios may need to use available cash to service their debt and may have difficulty in meeting their financial obligations.

A construction company is dependent on maintaining a sufficient level of cash to finance its daily operations. If adequate cash reserves are not maintained, the company must either borrow from a bank or default on its obligations. If the cost of borrowing was not included in a project budget, the interest charges on borrowed cash may erode anticipated profits. As illustrated in Figure 18.2, cash is invested in projects that then become accounts receivable. The receivables are collected and become cash. Some portions of the accounts receivable may be withheld as retainage by project owners until the completion of their projects. If the general contract specifies that retainage will be held by the owner, the general contractor usually includes the same provision in all subcontracts.

Because of the lag in receiving payment for accounts receivable (usually 30 days), the net cash flows on projects tend to be negative during the early phases of a project and positive during later phases. Predicting cash flow requirements involves conducting a cash flow analysis for the project. Project expenditures and revenues are estimated for each month, and the difference between revenues and expenditures are the cash flow requirements.

# Human resource management

Human resource management involves the design of the project management organization, recruitment and selection of new employees, management of subordinates' work performance, and professional development of subordinates. Organizational design involves identification of the basic project management organizational structure and determination of the specific responsibilities of each position. The result is an organization chart, which was discussed in Chapter 1, and a job description and set of job specifications for each position. The job description defines the major tasks and duties to be performed by the person selected for the position. The job specifications identify the knowledge, abilities, and skills needed to be able to perform the tasks and duties contained in the job description. An example job description and set of job specifications for an administrative assistant are shown in Box 18.3.

Both the job description and the job specifications are used when advertising for new applicants. The job description tells prospective applicants what the job entails, and the job

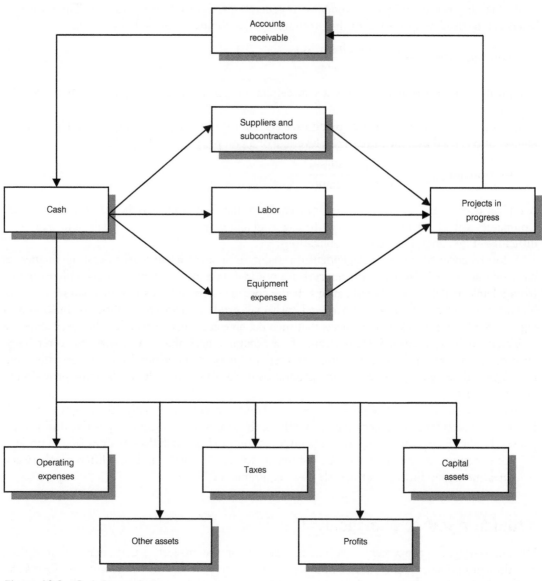

**Figure 18.2**   Cash flow cycle

specifications describe the knowledge, abilities, and skills an applicant must possess to be considered qualified for the position. The job description also is used by the project manager to establish an appropriate compensation rate for the position. Most construction firms have standard compensation schedules, which list all types of positions. Compensation decisions generally are based on the relative worth of the work to be performed and compensation levels for comparable positions in other construction firms. The objective is to make the position attractive to qualified individuals.

# Box 18.3   Job description and job specifications for administrative assistant

**Job Title:** Administrative Assistant

**Reports to:** Project Manager

**General Responsibilities:** Provides clerical and reception services for project field office. Assists project personnel in the preparation and processing of written, verbal, and electronic communications. Responsible for the operation and maintenance of project office equipment.

**Specific Responsibilities:**

1.   Provides reception services for the project office. Receives, records, and directs site visitors, telephone communications, electronic mail, internal distribution, parcel deliveries, and postal mail.

2    Assists the project manager in maintaining project files. Copies project documents as required by the project manager and company operating procedures.

3.   Provides data processing support to the project office. Inputs data as required into computer database management, spreadsheet, and other applications programs.

4.   Prepares final copies of letters, memorandums, standard forms, and other written communications using word processing programs.

5.   Coordinates maintenance of field office and office equipment to maintain a presentable, efficient, and functional environment. Ensures that all critical office equipment are operational at all times. Responsible for maintaining an operational stock of consumable office supplies.

**Required Knowledge, Skills, and Abilities:**

1.   Knowledge of construction terms and construction documents.

2.   Knowledge of the operation and user-level maintenance of standard office equipment.

3.   Knowledge of computer database management, word processing, and spreadsheet software applications.

4.   Skilled in keyboard operations for fast and accurate data input.

5.   Able to communicate effectively in person and over the telephone. Able to maintain a focus on customer service and to represent the company at the project in a professional, courteous manner.

6.   Able to organize, plan, and prioritize work efficiently. Must be able to work independently and productively with limited resources.

**Experience Requirements:** This position requires a minimum of five years experience in office administrative support functions, with a minimum of two years experience in construction field offices.

After the job descriptions and job specifications have been developed, the project manager determines with company executives which positions are to be filled with current employees of the construction firm and which are to be filled with outside applicants. Recruiting applicants for positions to be externally filled may be through an employment agency, newspaper advertising, Internet advertising, or university or technical school recruiting. Most construction firms require each applicant to submit a standard application form. In addition, résumés also are customarily required.

Once the applications have been received, the project manager must decide to whom to offer each position. Information used in making the decision generally is the information contained on the application form, any letters of recommendation provided, applicant's résumé, contacts with listed references, and applicant interviews. Most project managers want to select qualified individuals who are self-motivated and will relate well with other team members. The written documentation typically is used to screen the candidates and select the most qualified individuals. The top-ranked candidates should be interviewed to assess their qualities and qualifications for the position and to provide them information regarding the position. Interviews must be planned carefully to ensure each candidate is evaluated using the same criteria. Once new employees are selected, they must be oriented regarding company policies and the specific jobs that they are to perform as members of the project management team.

Performance management involves establishment of performance standards for each position and creation of a performance appraisal system. Performance standards typically are developed jointly by the employee and his or her supervisor. An example set of performance standards for an administrative assistant is shown in Box 18.4. A performance appraisal is a written evaluation of an employee's performance during a specific review period, typically 1 year. Forms similar to the one shown in Box 18.5 are used to document the evaluation. The appraisal is prepared by the employee's supervisor and describes how well the employee has performed during the review period. The appraisal should focus only on performance attributes that relate to organizational success. Appraisals often are used to determine compensation increases, annual bonuses, and employee training needs. The appraisal system works best when there is frequent feedback and mentoring provided by the supervisor to the employee throughout the review period, not just when the appraisal form is completed.

The last human resource management topic that we will discuss is employee development. There are two important aspects to employee development. The first is skill-enhancement training designed to improve employee performance in his or her current position. The second aspect is development training to provide the employee with the necessary skills to allow him or her to compete for higher-level positions within the company. Both aspects are important. The first step in creating an employee development program is to assess the needs of each employee for skill-enhancement training and for professional development. Next, specific training programs are either established or identified. Employees are sent to specific training programs, and their performance is monitored to ascertain the effectiveness of the individual training programs. An individual development plan should be established for each employee as a part of the appraisal process. This plan should be created jointly by the employee and his or her supervisor. Training needs should be prioritized, because most construction firms have limited training budgets.

## Box 18.4    Performance standard for administrative assistant

### Performance Standards for Administrative Assistant

Name of Employee:  Janet Smith

Supervisor:  Ted Jones

**Performance Standards:**

1.  Receive visitors, record their presence, and direct to appropriate office with no more than one error per month.

2.  Receive telephone communications, electronic mail, and written documents and direct to appropriate individual with no more than 2 errors per month.

3.  Maintain project files and misfile no more than 2 documents per month.

4.  Prepare final copies of written communication with no more than one error per week.

5.  Ensure reception area in project office is well organized and presents a professional image to visitors.

6.  Keep office reproduction and facsimile machines operational by contacting appropriate vendor to perform any needed repairs.

7.  Maintain a 30-day supply of consumable office supplies.

Employee Signature:  *Janet Smith*         Date:  August 24, 2016

Supervisor Signature:  *Ted Jones*         Date:  August 24, 2016

## Summary

Business development is the process of acquiring customers for the company. Marketing is the process of retaining desired customers and attracting new ones. Sales is obtaining the contracts for specific projects. Business development includes understanding the needs and expectations of current and prospective customers and ensuring that they are satisfied with completed projects.

Financial management is understanding the financial condition of the company and maintaining adequate cash flow on construction projects. Analysis of financial statements is used to assess the financial conditions of a company. Cash flow analysis is used to ensure that the company has sufficient cash resources to finance its operations.

## Box 18.5   Performance appraisal form

### Northwest Construction Company
### Annual Performance Appraisal

Name of Employee: _____

Name of Supervisor: _____

Evaluation Period: From _____ to _____

For each applicable performance area, mark the box that most closely reflects the employee's performance using the following scale:  1 = unacceptable; 2 = needs improvement; 3 = satisfactory; 4 = above average; and 5 = outstanding

| Performance Area | 1 | 2 | 3 | 4 | 5 |
|---|---|---|---|---|---|
| Ability to make job-related decisions | | | | | |
| Accepts responsibility | | | | | |
| Attendance | | | | | |
| Attitude | | | | | |
| Cooperation | | | | | |
| Dependability | | | | | |
| Effective under stress | | | | | |
| Initiative | | | | | |
| Leadership | | | | | |
| Operation and care of equipment | | | | | |
| Quality of work | | | | | |
| Safety practices | | | | | |
| Technical abilities | | | | | |

**Job Strengths:** _____

_____

**Areas for Improvement:** _____

_____

**Training Needed:** _____

_____

**Supervisor Comments:** _____

_____

**Employee Comments:** _____

_____

Employee Signature: _____   Date: _____

Supervisor Signature: _____   Date: _____

Human resources management involves the design of the project management organization, recruitment and selection of new employees, management of subordinates' work performance, and professional development of subordinates. Once the organizational structure has been developed for the project office, a job description and job specifications are prepared for each position. These are used in recruiting new employees and developing performance standards. Performance appraisals are used to evaluate and document subordinate performance and identify any training needed.

## Review questions

1. What are the primary responsibilities of the business development staff of a construction company?
2. What is the project team's role in winning a construction contract awarded using a negotiated procurement process?
3. Why do general contractors typically require prospective subcontractors to submit current financial statements during the prequalification process?
4. Why is a liquidity assessment important when analyzing a prospective subcontractor's balance sheet?
5. What is the process for determining the monthly cash flow requirements on a construction project?
6. What is a job description, and what is it used for?
7. What are job specifications, and what are they used for?
8. What type of written documentation would you use when selecting applicants for interviews?
9. What are performance standards, and what are they used for?
10. Who prepares the performance appraisal, and what is it used for?
11. How would you prepare an individual development plan for a subordinate employee?

## Exercises

1. Determine the current ratio, return-on-assets ratio, and the debt-to-equity ratio for Eastside Excavating using the financial statements shown in Box 18.1 and Box 18.2. What is your assessment of the financial condition of the company?
2. Prepare a job description for a field engineer.
3. Prepare a performance standard for a field engineer.

# 19 Construction project leadership

## Introduction

What does construction leadership mean, and who is a construction leader? Dictionaries do not define the specific term *construction leadership*, so we start with parts and pieces.

*Construction*: The arrangement and connection...The process, art, or manner of constructing...A creation that is put together out of separate pieces of often disparate materials.

*Leader*: Someone that leads...Someone that ranks first...A person who directs a force or unit...A person who has commanding authority or influence...The principal officer...One that exercises paramount but responsible authority over an organization...The principal member endowed to govern with a minimum of formal restraints.

*Leadership*: The office or position of a leader... The quality of a leader... Capacity to lead.

So the answer should be as simple as adding these three together. However, if that were the case, this section of the chapter would be the summary rather than the introduction. The fact is that there is not much written specific to construction leadership, let alone any accepted definition. There are literally hundreds if not thousands of books and academic articles written on the topic of leadership, but most of them focus on Fortune 500 company chief executive officers (CEOs), military and political leaders, athletic coaches, and religious leaders. There is not any book dedicated to the specific topic of construction leadership. Even most of the academic articles with "construction leadership" in their titles really do not focus on the construction industry or construction personnel. For the basis of this chapter, we conducted approximately 300 surveys and interviews with construction leaders and researched and developed more than 100 case studies on famous construction projects and famous construction leaders.

Beginning with the basic concepts above, adding definitions from several other researchers and authors of leadership, incorporating our case studies, and summarizing our surveys and interviews, we developed the following definition of a construction leader as a person who:

- guides others in the construction of a project through knowledge of work processes and thorough planning;
- has authority or influence over the construction process due to superior experience or position and is responsible for its outcome;
- creatively inspires and enables others to perform construction tasks; and
- has the skills, abilities, authority, power, and capacity necessary to guide, motivate, and influence the craftsmen in a way minimizing conflict and resulting in a successful construction project.

Our goal of researching the topic of construction leadership has been: Do traditional leadership research and theories apply to construction? If not, why not? If so, how does it? And if it does, why has there been so little written specific to construction leadership? The easy answers are, of course, it all applies exactly. The reality is that the construction industry is unique. It is not the same as automobile manufacturing. We will never build the same building on the same site with the same set of subcontractors and suppliers and craftsmen at the same time of year and so on. Every project is a unique project in some regard. So, therefore, our first conclusion to the foregoing question was no, traditional theory does not apply to the construction industry. Similar to previous chapters, our focus for the topic of construction leadership will be at the jobsite project level focusing on project managers and superintendents.

## History of leadership theories

Research into what makes a great leader has been going on for literally hundreds of years. There has never been an accepted theory or a consensus, so the research and articles and books continue today. We will look at just a few of the theories.

## The Great Man Theory

Historically, it was felt that great leaders were born into the role. These people not only affected history, but they created it. Examples included Moses, Winston Churchill, kings of countries, Thomas Jefferson, Gandhi, and others. These people were not specifically chosen for their leadership potential or took a class or were trained for the job; it was felt many inherited the position or the skills from their ancestors. In many cases, these types of leaders are remembered for their leadership skills in times of crisis.

## The Trait Theory

Similar to the Great Man Theory, this theory assumes leaders are born with certain traits that others don't have and, by studying those traits, we should be able to predict who will be the next or the best leader (i.e., choose the leader of tomorrow from a lineup). Testing showed that leaders outperform followers, so they would choose the person with the most potential and work with them. If a person was proven a leader in one category, it was assumed that he or she could be a leader in other categories. This theory fell out of favor due to consideration of personal behaviors. In addition, a particular leadership situation must be considered.

## Situational leadership

This theory disagrees with the Great Man and Trait Theories. Leadership is more a case of being in the "right place at the right time." The proper choice of leader and leadership style depends upon many factors, including bureaucracy, organization, interpersonal behaviors, autocracy between superiors and subordinates, job design, and integration of individual and organizational goals. Organization changes will also change leadership requirements and the proper leader choices.

## Adaptive theory

The leader must not only know him- or herself but know his or her people and adapt the right approach to the right situation. It implies that the leader has the ability to be different types of people in different situations, sometimes behaving one way with one foreman on a lump sum project, for example, and another way with a subcontractor PM on a negotiated project.

Many of the more current theorists combine several of the aforementioned theories and others not listed and come up with a complex resolution which basically concludes with "it depends."

Most of these studies include analysis of specific accepted leaders as case studies, or surveys, but none of them use construction leaders. Why is that? Is one born to be a construction leader? Do they inherit that trait from their parents, who may have been construction leaders? Can they be trained on the job through supervisory training programs, such as offered by the associated General Contractors or other trade and industry organizations, or through collegiate construction management (CM) programs? Can a construction superintendent "adapt" to his or her situation with a given set of subcontractors, craft workers, client, architect, design documents, site, and the like?

## Leadership potential

Are construction jobsite leaders born into their role or do they grow into it? Can *any* apprentice carpenter become a journeyman, foreman, and superintendent? Can *any* CM graduate who starts out as a field or project engineer become a project manager (PM), senior PM, and eventually a CEO or owner of a construction firm? Some of the more famous construction leaders we studied included:

- Robert Stevenson, Bell Rock Lighthouse;
- John, Washington, and Emily Roebling, Brooklyn Bridge;
- Joseph Strauss, Golden Gate Bridge;
- Joseph Banks and Rufus Woods, Grand Coulee Dam;
- Frank Crowe and Steven Bechtel, Hoover Dam; and
- Peter Kiewit, Peter Kiewit and Sons.

Are we all capable of designing, constructing, or master-building these types of projects? The simple answer is no; there are limits on leadership potential in the construction industry. There are many foremen who neither could nor would want to become superintendents. There are project engineers who are comfortable in developing requests for information and submittals and have no interest facing an owner with a pay request or a change order as would be expected of the project manager. Although as discussed in the previous section, it is difficult to measure or explain why, but a construction leader has some inherent talent. That talent needs to be further educated, grown, and nurtured in order to become a successful leader.

There are many factors that contribute to determining an individual's leadership abilities. Leadership is something that someone can be born with, to some extent, yet it is the topic of countless seminars and self-help books as a skill that can be learned and refined over time. In the construction industry, it takes some of both to be an effective leader. Leaders in construction are individuals who are born with a self-motivated drive to produce work that excels above and beyond their competitors, making them stand out in a crowd. These leaders take their skills and apply them to the work they are passionate about, learning from their mistakes and never missing an opportunity to improve themselves and those they work with. Regardless of the level of their passion, however, a person's leadership ability is the determining factor for their effectiveness. This concept seems obvious; highly effective leaders in the past have had a great ability to lead. If their leadership abilities were lower, then their effectiveness would have duly suffered, and those they were leading would have found a more effective leader as a result.

The construction industry is one that is full of leaders. This is obvious; without effective leaders, there would be fewer completed projects and more litigation and setbacks to progress. In this industry, ineffective project managers and superintendents are quickly replaced by those more qualified to handle the task at hand. It takes true leaders to be able to take an idea and translate it into a physical and often habitable element of the built environment. Without a heightened level of leadership ability, these individuals would lose their effectiveness as a leader. Leadership ability is the determining factor to a person's effectiveness. A construction leader needs to be willing to spend the time to learn how to be a better leader. Becoming an effective construction leader doesn't happen overnight; it takes time to learn about the construction business and understand what it takes. A construction leader needs to not only know how to manage a project but to know how to lead a project. Leadership growth is a lifelong process, and this applies to the construction industry maybe more than most others.

## Profile of a construction leader

What does a construction leader look like? Is he or she tall or short, thin or stocky, male or female? The answer is yes. How do construction leaders behave? Are their qualities and skills similar to other Fortune 500 or military leaders? Our first premise was no: They are different, but a lot of our survey results indicated that, especially at the CEO level of larger construction firms, many of the leadership traits were similar. What we found at the project management and supervision levels were different, and that is because construction projects are unique and not assembly-line.

Many of the books and articles on traditional leadership list characteristics, behaviors, and traits common among leaders. Many of the early traditional theories use lists to sort through groups of individuals to target who the next CEO or military general or athletic coach should be and provide them with the proper training so that they could fulfill their destiny. The list of

---

## Box 19.1   Construction leadership traits

| | |
|---|---|
| Accountable/Ethical | Influential/Respect |
| Adaptive/Flexible | Knowledge |
| Charismatic | Motivational |
| Communicator | Trust |
| Confident/Assertive | Vision |

---

potential traits is quite long, but our research of construction leaders yielded the top ten listed in Box 19.1.

A lot of leadership research and debate also focuses on the differences between *management* and *leadership*. Some indicate they are entirely two different types of people and roles. Others feel they are the same. However, a lot of the current thought is that there is a significant overlap and both traits are needed to be successful as is shown in Figure 19.1. In our research, we found there is some distinction with positions and titles. A mid-sized construction company PM actually has very few "employees" working for them, and most of their management skills are related to document flow and subcontractor coordination. CEOs of larger construction firms were at one time successful project managers who had additional leadership skills. Superintendents who are pushing the field work wear both hats; but similar to the PM, they have the ability to enforce a contract with respect to their subcontractors and a fairly large labor pool to choose from to keep their craft workers motivated. Again, many of the current publications indicate that the position doesn't define a leader, but it would be difficult to believe that the apprentice carpenter or the field engineer on a project is actually leading it to success.

## Construction leadership myths

The previous profile was that construction leaders are all tall white males who played collegiate football, scream loudly, and are frequent users of profanity. A few maybe had these characteristics, especially 50 or so years ago, but not so today. What follows are some specific examples of realities about construction leadership myths:

- The construction industry is adopting a more business-style approach, where people skills and interpersonal relationships are highly valued. One does not have to be physically intimidating and vociferous. Respect is not earned by a loud voice but is based on past successes.
- With today's fast-tracked projects and very tight budgets, a construction leader has to implement the latest technology to stay on schedule; change is part of today's construction process.
- Partnering relationships and trust between companies and individuals are paramount to foster collaboration and avoid adversarial situations. Construction is not all controlled by exact

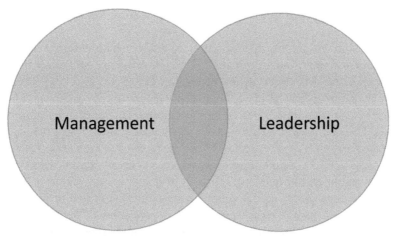

**Figure 19.1**    Management vs leadership

contract language. As project owners adopt alternative project delivery methods, collaborative approaches to construction execution become more critical.

- In construction, strong technical skills are essential for a leader. Even if one is naturally a good communicator, technical knowledge is not something you are born with. A leader in construction works himself or herself up through the ranks by on-the-job-training.
- Although one might be attracted by a leader's charisma, workers on the jobsite tend to also look for trust and reputation in a PM and superintendent.
- A successful leader in construction must share his or her vision with other members of the team and organization and in return consider their ideas. Leadership today is about focusing on people's talents, enthusiasm, and earnest intent to achieve common goals; it is not solely driven by a paycheck on Friday.
- What it takes to truly build a building are technological, contracting, risk management, cash flow, and several other tools and skills. Today's constructor is likely college-educated and may have a professional license.
- There are many excellent women field supervisors, project managers, and superintendents, as well as CEOs. Gender barriers are being broken down in this industry, and the number of women and minority leaders in construction is growing daily.
- Construction leaders are on the job 24–7; they are always planning the next move. It is not a Monday through Friday, 8–5 job.
- Good people and communication skills are essential to enable construction leaders to network with potential clients and to form collaborative teams for construction projects.

## Teamwork

Can a construction tradesman build a project all by him- or herself? This author's father just about could. He was a master carpenter and as close to a modern-day master builder as I have observed, but his focus was on speculative and custom home building; one at a time. It is fair to say that a commercial building in the $5 to $50 million range could not be built with just one person; a team is needed. Leaders need followers. How do they acquire followers they can trust to help them achieve success?

Internal relationships in a company are easier to analyze than external ones, because it is a more controlled environment. Often seen at a company's office are employees that share the same traits as their leader, and it is clearer when you go up the ranks. The field situation is different than the office; the construction team of superintendent, foreman, craftsmen, and laborers often don't work together for more than a year. The superintendent tries to hold on to his key foremen for the next job. The relationship between superintendents and their foremen can be very close, especially in an economic crisis where many foremen agree to work as craftsmen and take a pay cut in order to stick with their superintendent, which is a great example of attraction and even sacrifice. The leadership attribute of attraction is difficult to notice clearly because the teams are so dynamic. The concept of "who's in your inner-circle" as shown in Figure 19.2 applies to construction superintendents as well as to project managers and CEOs.

Construction leaders add value to their followers by including them in the bigger picture. Close allies are developed both internally, such as relations between foremen and superintendents, and externally with repeat projects between general contractors and their clients. Buildings are set on firm concrete foundations, just as construction leaders use a foundation of close connections and solid ideals to help achieve their goal of repeat teams and successful projects.

## Respect

The title of "leader" is usually not enough; although it helps. In the construction industry. carrying the title on the jobsite of superintendent or PM is certainly a plus. However, the leader still needs to earn the respect of his or her followers. How do they do that? Construction workers, especially subcontractors, will quickly determine whether the PM or superintendent can be led astray. A person's reputation and their background are a plus.

In the construction field, respect is critical between leaders and followers. The construction industry is constantly changing, and the faces of the leaders on every project change with it. Part of the leadership development process for a construction leader is earning respect, whether it is in the office or on the jobsite. Sometimes, a young, inexperienced PM is placed in a leadership position and is expected to lead superintendents, subcontractors, or other personnel who have more experience. It is critical that the PM earn the respect of the superintendent and others in order to achieve success on the project and accomplishment of goals and team development. Since knowledge, skills, and abilities are developed over time, one of the ways that a PM can *earn respect* when he or she lacks experience *is to show respect* to those more experienced who are expected to follow their leadership. The same can be said for trust. A leader who demonstrates a sincere interest in learning and developing these skills—and trusts and shows respect to those who already have the skills—will earn their trust and respect in return, even if he or she is a "green" manager. This is respect in 360 degrees as depicted in Figure 19.3.

Hiring from outside rarely works in construction; leaders are usually promoted from within the company. We observe in construction a lot of the following:

• Owners and architects, if possible, employ general contractors with whom they have had prior success.

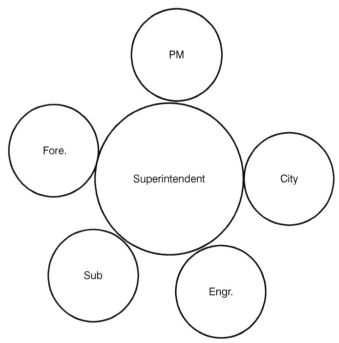

**Figure 19.2**  Superintendent's inner-circle

- General contractors lean toward subcontractors who have proven to provide a quality project.
- Project managers have project engineers (PEs) whom they bring from job to job, as do superintendents who bring along foremen and foremen who bring along journeymen with whom they work well.

However, we have short memories in construction as well. All it usually takes is one mistake or one bad project and the string of referrals is broken. One sure way to earn respect is to give it to others. This author has often told younger PMs or PEs that they will need a favor from every subcontractor and every foreman on the project sooner or later, and if they mistreat those people, that favor will not be forthcoming. You may need to have a couple of carpenters stay an hour or so late one day finishing a task or ask a subcontractor to work a Saturday to make up for someone else's being behind schedule or a back charge or change order forgiven. Even borrowing a forklift may be difficult if respect was not granted to those you work with on prior occasions.

One of the biggest challenges for leaders in construction is to lead people over whom they have no authority. A field engineer may be placed in charge of a subcontractor but does not have direct authority, hire and fire privileges, or control of the subcontractor's pay. A third-party agency owner's representative is responsible for coordinating the interactions between the owner, architect, and contractor but often has no contractual authority over any of them. There are numerous examples within construction of leaders needing to influence a number of people to follow a plan or achieve a task and having only their charisma and communication skills to leverage their agenda. A good leader has to have the ability to gain followers and influence them to work toward a common goal in any circumstance.

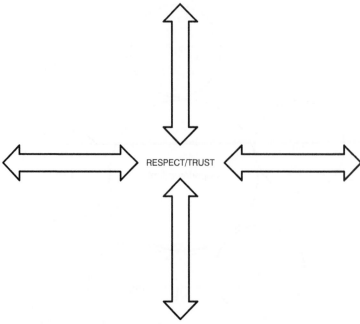

RESPECT/TRUST

**Figure 19.3**    360 degree view

## Craft motivation

Motivation is one of those terms that are overused. "Lack of motivation" or "how to motivate" are common phrases in the academic and business worlds. Motivation is a difficult term to define or quantify and even more difficult to resolve. If it were easy, why would there be so many papers and workshops on this one particular topic? This author has taught motivational seminars to contractor trade associations and knows first-hand how difficult a concept it is to get one's arms around. Many leadership authors write on how leaders need to motivate their followers—easier said (or written) than done.

What do successful Fortune 500 leaders use to motivate their followers? What motivates a construction leader? In construction, we ask our employees to work in excessive heat, cold, rain, and sometimes dangerous conditions. Accidents happen, and sometimes workers lose their lives on construction projects, although jobsite working conditions have certainly improved in the last 50 years. The old rule of thumb of one death per million dollars of construction, which Frank Strauss took as a challenge and beat on the Golden Gate Bridge, certainly doesn't apply today. Zero deaths is not only a goal, it is a requirement, and zero accidents is a worthwhile and often achievable goal on many construction projects.

The answer to the question of motivating construction craftsmen is not just paying them a fair wage every Friday, although many construction supervisors still feel that way. Supervisors ranked wages, job security, and personal growth at the top of their expected list, but our surveys showed these were mid-choices with the employees. Employees chose appreciation, involvement, and sympathetic help at the top of their list, the three items the superintendents felt would be at the bottom as is shown in Box 19.2.

## Box 19.2    Field motivational factors

| Motivational Factors | Supervisors' Ranking for Employees | Employees' Ranking |
|---|---|---|
| Full Appreciation | 8 | 1 |
| Feeling of Being in on Things | 10 | 2 |
| Sympathetic Help | 9 | 3 |
| Job Security | 2 | 4 |
| Wages | 1 | 5 |
| Interesting Work | 5 | 6 |
| Personal Growth | 3 | 7 |
| Loyalty | 6 | 8 |
| Working Conditions | 4 | 9 |
| Tactical Discipline | 7 | 10 |

## Vision

Do leaders have a crystal ball or, in the case for construction, a transit or a laser or total-station that tells them the right course of action to take? Can they see into the future? No, of course not, but many authors and researchers believe that a leader takes a long-term view of the company and does not focus on the details. How does the ability to see the big picture affect construction leadership? Again, with the CEO of a larger firm, they likely are taking the long view of company success gained through client and community relationships and developing their successors. However, on a project level, the superintendent and PM also have a vision, and that is what the project will look like when it is done and what the team needs to do today, next week, and next month in order to bring it in on time and on schedule. In recent years, with the advent of computer-aided design and building information modeling through software programs such as Revit and Sketch-up, the ability to visualize the finished project has become easier, but it still requires the superintendent and PM to be able to explain that view to the craftsmen and the subcontractors in the field.

Planning appropriately is synonymous with terms such as *planning* and *scheduling* and *preconstruction planning*, topics introduced in Chapters 4 and 5. Although none of us actually has a crystal ball to look into the future, planning implies doing whatever is possible to try to anticipate the outcome of one's actions, or inactions, and preparing accordingly. The PM and superintendent cannot expect their employees to act when they themselves will not. Therefore, especially in construction, a leader must lead by example. On one rainy day in the Pacific

Northwest, this author put on his rain gear and boots and walked an entire project and talked with all the craftsmen. Their response to the owner's representative who could have been in the trailer drinking coffee was "glad to see you out here Len, we appreciate it". Some may say you are born with intuitive skills, and some may have more inherent ability at this than do others, but in construction, knowledge and experience are powerful traits. If the project team has built many projects and they have planned this one to the best of their abilities, vision—or intuitive skills—will assist when quick and decisive actions are necessary. Construction leaders truly walk the talk.

## Winning

Some owners and architects might feel that it is the goal of the contractors to win and achieve a profit at all costs. Some in our industry may take that perspective, but is it true with all firms? Contractors want to win the bid or be the successful proposing contractor or win the negotiation on a change order or a claim, but will they do it at any cost? Will they take a chance to lose a relationship or cross the ethics line to achieve success? Most will tell you no, but unfortunately, there are a few bad apples in every barrel.

Construction is a competitive industry, and leaders find ways to win, regardless of the economy, but a victory at all costs is not the type of victory they are looking for. Quality and safety continue to reign paramount in construction. Regardless of the size or complexity of a project, being able to set priorities is crucial. Leaders guide those who are involved with the project through its conclusion. In order to achieve these goals effectively, it is critical to set priorities for the present as well as looking ahead to understand what needs to be accomplished next and who will accomplish these tasks in a timely and cost-efficient manner. Without priorities and leaders to set them, projects would be disorganized. Good leaders are a necessity for success. Great leaders know that sacrifice is necessary. They understand that no one can become a true leader without sacrificing their personal interests: time, money, relationship, family life, or even their health. When the time comes, leaders are not afraid and are more than willing to sacrifice for their vision and dream. Sacrifices are the essence of true construction leadership. In order to be successful and "win," or even just keep the present state, the PM and superintendent must make sacrifices based on situations every day. Furthermore, when the leader wants to be at the highest level in his or her field, the leader must give up many more things, at the same time incurring more responsibilities. Sacrifices are an ongoing process, not a one-time occurrence but often necessary to be successful.

## Leaders develop leaders

Construction is a unique industry in many ways, as has been discussed. We seek and achieve immediate positive feedback. We are fortunate to be able to see the results of our day's or year's work by viewing a completed construction project, but a more important test of a construction leader is whom he or she has trained to take their place. Who will continue their practice of building quality projects on time and on budget all the while maintaining a safe workplace?

A lot of people are leaders. Many were born with more leadership skills than others. Many others developed and continue to develop their leadership skills on the job, especially in the construction industry. However, being born with leadership skills and taking a seminar or two or reading a couple of books doesn't qualify one to be a "great" leader. Leaders need followers, and it is relatively simple for leaders to lead those who want to be led, but finding and leading leaders and developing future leaders is what really defines a great leader. Leaders expand their organizations by developing leaders, not by developing followers. Developing followers grows organizations only one person at a time, whereas developing leaders makes the organization not only attract leaders but also additional followers that follow their leaders. Developing leaders makes the business grow by teams instead of by individuals.

If a PM or superintendent empowers his or her project engineers and foremen and helps them to develop so that they become capable of taking over their supervisor's job, they will make themselves indispensable to their organizations. This is easy to see in construction.

- Foremen develop journeymen to become future foremen.
- Superintendents develop foremen to become future superintendents.
- Project managers develop project engineers to become future project managers.
- Corporate officers develop project managers and senior project managers to become corporate officers and eventual owners of the construction firm.

## Summary

There is very little written specific to the study of construction leadership. Many of the previous leadership theories have evolved into a situational model, basically "it depends." Many of us have potential leadership capabilities, but it takes a combination of natural talent and acquired or learned skills to be an affective construction leader. Construction leaders need a variety of tools in their toolboxes, including the ability to be both a manager and a leader. Many stereotypes of construction "bosses" or leaders have thankfully evolved into project managers and superintendents who are sensitive about those whom they work for, including clients and architects, and those they supervise at the project level. Traits such as respect and trust function in a 360-degree fashion in that they must be given to be received. Top motivational factors for construction trades include being appreciated and included in the big picture, not just wages. Through thorough pre-project planning, the project manager's and superintendent's vision of a successful project assists with leading their team in the field. Their legacy lies not only with a happy client and architect but also with qualified and trained employees who can someday lead their own project teams.

## Review questions

1. Why do you feel there have not been any books written specific to construction leadership?
2. Who is one of your favorite famous political, military, or sports leaders? Why them?
3. Of all of the "traditional" leadership theories, which one applies to construction leaders the best?
4. Other than our list of construction leadership myths, what stereotype do you know of?
5. What did you dislike about a former or current boss, and what did you like the most about a former or current boss?
6. Was your favorite boss a manager or a leader or both?

7. Of our top 10 leadership traits, which one do you feel should rank first?
8. What motivates you as a student? What motivates you as an employee? What do you think motivates your favorite boss?
9. Are you a leader? If not now, what would it take for you to become a leader?

## Exercises

1. Develop an inner-circle diagram similar to Figure 19.2 above for (a) project manager, (b) CEO, (c) project engineer, (d) the client, and/or (e) yourself.
2. Looking back at Figure 19.1, how do you see the circles overlapping with respect to (a) construction CEO, (b) project engineer, (c) structural engineer, (d) journeyman carpenter, and/or (e) yourself?
3. Our example list of construction leaders is very brief. Whom do you feel should be added to a list of design and construction leaders?

# Glossary

**Active quality management program:**  process that anticipates and prevents quality control problems rather than just responding to and correcting deficiencies

**Activity duration:**  the estimated length of time required to complete an activity

**Addenda:**  additions to or changes in bid documents issued prior to bid and contract award

**Additive alternates:**  alternates that add to the base bid if selected by the owner

**Agency construction management delivery method:**  a delivery method in which the owner has three contracts: one with the architect, one with the general contractor, and one with the construction manager. The construction manager acts as the owner's agent but has no contractual authority over the architect or the general contractor

**Agreement:**  a document that sets forth the provisions, responsibilities, and the obligations of parties to a contract. Standard forms of agreement are available from professional organizations.

**Allowance:**  an amount stated in the contract for inclusion in the contract sum to cover the cost of prescribed items, the full description of which is not known at the time of bidding. The actual cost of such items is determined by the contractor at the time of selection by the architect or owner, and the total contract amount is adjusted accordingly.

**Alternates:**  selected items of work for which bidders are required to provide prices

**Alternative dispute resolution:**  a method of resolving disagreements other than by litigation

**American Institute of Architects:**  a national association that promotes the practice of architecture and publishes many standard contract forms used in the construction industry

**Application for payment:**  see *payment request*

**Arbitration:**  method of dispute resolution in which an arbitrator or a panel of arbitrators evaluates the arguments of the respective parties and renders a decision

**Arrow diagramming method:**  scheduling technique that uses arrows to depict activities and nodes to depict events or dates

**As-built drawings:**  contractor-corrected construction drawings depicting actual dimensions, elevations, and conditions of in-place constructed work

**As-built estimate:**   cost estimate in which actual costs incurred are applied to the quantities installed to develop actual unit prices and productivity rates

**As-built schedule:**   marked-up, detailed schedule depicting actual start and completion dates, durations, deliveries, and restraint activities

**Associated General Contractors of America:**   a national trade association primarily made up of construction firms and construction industry professionals

**Automobile insurance:**   protects the contractor against claims from another party for bodily injury or property damage caused by contractor-owned, -leased, or -rented automobiles and equipment operated over the highway

**Back charge:**   general contractor charge against a subcontractor for work the general contractor performed on behalf of the subcontractor

**Balance sheet:**   a financial statement that represents the financial condition of a company as of the date of the balanced sheet in terms of assets, liabilities, and owners' equity

**Bar chart schedule:**   time-dependent schedule system without nodes that may or may not include restraint lines

**Bid bond:**   a surety instrument that guarantees that the contractor, if awarded the contract, will enter into a binding contract for the price bid and provide all required bonds

**Bid security:**   money placed in escrow, a cashier's check, or bid bond offered as assurance to an owner that the bid is valid and that the bidder will enter into a contract for that price

**Bid shopping:**   unethical general contractor activity of sharing subcontractor bid values with the subcontractor's competitors in order to drive down prices

**Blanket purchase order:**   see *open purchase order*

**Bridging delivery method:**   a hybrid of the traditional and the design-build delivery methods. The owner contracts with a design firm for the preparation of partial design documents, then selects a design-build firm to complete the design and construct the project.

**Builder's risk insurance:**   protects the contractor in the event that the project is damaged or destroyed while under construction

**Building information models:**   computer design software involving multi-discipline three-dimension overlays improving constructability and reducing change orders

**Build-operate-transfer delivery method:**   a delivery method in which a single contractor is responsible for financing the design and construction of a project and is paid an annual fee to operate the completed project for a period of time, such as 30 years

**Buyout:**   the process of awarding subcontracts and issuing purchase orders for materials and equipment

**Buyout log:**   a project management document that is used for planning and tracking the buyout process

**Cash flow curve:**   a plot of the estimated value of work to be completed each month during the construction of a project

**Cash-loaded schedule:**   a schedule in which the value of each activity is distributed across the activity, and monthly costs are summed to produce a cash flow curve

**Certificate of insurance:**   a document issued by an authorized representative of an insurance company stating the types, amounts, and effective dates of insurance for a designated insured

**Certificate of occupancy:**   a certificate issued by the city or municipality indicating that the completed project has been inspected and meets all code requirements

**Certificate of substantial completion:**   a certificate signed by the owner, architect, and contractor indicating the date that substantial completion was achieved

**Change order:**   modifications to contract documents made after contract award that incorporate changes in scope and adjustments in contract price and time

**Change order proposal:**   a request for a change order submitted to the owner by the contractor, or a proposed change sent to the contractor by the owner requesting pricing data

**Change order proposal log:**   a log listing all change order proposals indicating dates of initiation, approval, and incorporation as final change orders

**Claim:**   an unresolved request for a change order

**Close-out:**   the process of completing all construction and paperwork required to complete the project and close-out the contract

**Close-out log:**   a list of all close-out tasks that is used to manage project close-out

**Commissioning:**   a process of ensuring that all equipment is working properly and that operators are trained in equipment use

**Conceptual cost estimate:**   cost estimates developed using incomplete project documentation

**Conditional lien release:**   a lien release that indicates that the issuer gives up his or her lien rights on the condition that the money requested is paid

**ConsensusDocs:**   family of contract documents developed and endorsed by a consortium of 40 construction professional organizations

**Constructability analysis:**   an evaluation of preferred and alternative materials and construction methods

**Construction change directive:**   a directive issued by the owner to the contractor to proceed with the described change order

**Construction manager-at-risk delivery method:**   a delivery method in which the owner has two contracts: one with the architect and one with the construction manager/general contractor. The general contractor usually is hired early in the design process to perform preconstruction services. Once the design is completed, the construction manager/general contractor constructs the project.

**Construction manager/general contractor delivery method:**   see *construction manager-at-risk delivery method*

**Construction Specifications Institute:**   the professional organization that developed the original 16-division MasterFormat that is used to organize the technical specifications. Today's MasterFormat includes 49 divisions

**Contract:**   a legally enforceable agreement between two parties

**Contract time: also known as project time:**   The period of time allotted in the contract documents for the contractor to achieve substantial completion

**Coordination Drawings:**   multi-discipline design drawings which include overlays of mechanical, electrical, and plumbing systems with the goal of improving constructability and reducing change orders

**Corrected estimate:**   estimate that is adjusted based on buyout costs

**Cost codes:**   codes established in the firm's accounting system that are used for recording specific types of costs

**Cost estimating:**   process of preparing the best educated anticipated cost of a project given the parameters available

**Cost-plus-award-fee contract:**    a type of cost-plus contract in which the contractor's fee has two components; a fixed component and an award component based on the contractor's performance. The award component is decided at periodic intervals, for example quarterly, based on an owner-established set of criteria. Criteria might include cost control, quality of construction, safety performance, and meeting an agreed schedule.

**Cost-plus contract:**    a contract in which the contractor is reimbursed for stipulated direct and indirect costs associated with the construction of a project and is paid a fee to cover profit and company overhead

**Cost-plus contract with guaranteed maximum price:**    a cost-plus contract in which the contractor agrees to bear any construction costs that exceed the guaranteed maximum price unless the project scope of work is increased

**Cost-plus-fixed-fee contract:**    a cost-plus contract in which the contractor is guaranteed a fixed fee irrespective of the actual construction costs

**Cost-plus-incentive-fee contract:**    a cost-plus contract in which the contractor's fee is based on measurable incentives, such as actual construction cost or construction time. Higher fees are paid for lower construction costs and shorter project durations..

**Cost-plus-percentage-fee contract:**    a cost-plus contract in which the contractor's fee is a percentage of the actual construction costs

**Cost-reimbursable contract:**    a contract in which the contractor is reimbursed stipulated direct and indirect costs associated with the construction of a project. The contractor may or may not receive an additional fee to cover profit and company overhead.

**Craftspeople:**    non-managerial field labor force who construct the work, such as carpenters and electricians

**Critical path:**    the sequence of activities on a network schedule that determine the overall project duration

**Daily job diary: also known as daily journal or daily report:**    A daily report prepared by the superintendent that documents important daily events including weather, visitors, work activities, deliveries, and any problems

**Davis-Bacon wage rates:**    prevailing wage rates determined by the U.S. Department of Labor that must be met or exceeded by contractors and subcontractors on federally funded construction projects

**Deductive alternates:**    alternates that subtract from the base bid if selected by the owner

**Design-build delivery method:**    a delivery method in which the owner hires a single contractor who designs and constructs the project

**Design-build-operate delivery method:**    a delivery method in which the contractor designs the project, constructs it, and operates it for a period of time, for example, 20 years

**Detailed cost estimate:**    extensive estimate based on definitive design documents. Includes separate labor, material, equipment, and subcontractor quantities. Unit prices are applied material quantity take-offs for every item of work.

**Direct construction costs:**    labor, material, equipment, and subcontractor costs for the contractor, exclusive of any mark-ups

**Dispute:**    a contract claim between the owner and the general contractor that has not been resolved

**Dispute resolution board:** a panel of experts selected for a project to make recommendations regarding resolution of disputes brought before it

**Earned value:** a technique for determining the estimated or budgeted value of the work completed to date and comparing it with the actual cost of the work completed; used to determine the cost and schedule status of an activity or the entire project

**Eichleay Formula:** a complicated method of potentially recovering home office overhead and lost fee potential usually associated with claims involving time extensions

**Eighty–twenty rule:** on most projects, about 80 percent of the costs or schedule durations are included in 20 percent of the work items.

**Electronic mail:** Internet tool for sending communications and attached documents

**Equipment floater insurance:** protects the contractor against financial loss due to physical damage to equipment from named perils or all risks and theft

**Errors and omissions insurance: also known as professional liability insurance:** Protects design professionals from financial loss resulting from claims for damages sustained by others as a result of negligence, errors, or omissions in the performance of professional services

**Estimate schedule:** management document used to plan and forecast the activities and durations associated with preparing the cost estimate; not a construction schedule

**Exhibits:** important documents that are attached to a contract, such as a summary cost estimate, schedule, and document list

**Expediting:** process of monitoring and actively ensuring vendor's compliance with the purchase order requirements

**Expediting log:** a spreadsheet used to track material delivery requirements and commitments

**Experience modification rating:** a factor, unique to a construction firm, that reflects the company's past claims history; this factor is used to increase or decrease the company's worker's compensation insurance premium rates.

**Fast-track construction: also known as phased construction:** Overlapping design and construction activities so that some are performed in parallel rather than in series; allows construction to begin while the design is being completed

**Fee:** contractor's income after direct project and job site general conditions are subtracted; generally includes home office overhead costs and profit

**Field engineer:** similar to the project engineer except with less experience and responsibilities; may assist the superintendent with technical office functions

**Filing system:** organized system for storage and retrieval of project documents

**Final completion:** the stage of construction when all work required by the contract has been completed

**Final inspection:** final review of project by owner and architect to determine whether final completion has been achieved

**Final lien release:** a lien release issued by the contractor to the owner or by a subcontractor to the general contractor at the completion of a project indicating that all payments have been made and that no liens will be placed on the completed project

**Float:** the flexibility available to schedule activities not on the critical path without delaying the overall completion of the project

**Foreman:** direct supervisor of craft labor on a project

**Free on board:**   an item whose quoted price includes delivery at the point specified; any additional shipping costs to be paid by the purchaser of the item

**Front loading:**   a tactic used by a contractor to place an artificially high value on early activities in the schedule of values to improve cash flows

**General conditions:**   a part of the construction contract that contains a set of operating procedures that the owner typically uses on all projects; they describe the relationship between the owner and the contractor, the authority of the owner's representatives or agents, and the terms of the contract.

**General contractor:**   the party to a construction contract who agrees to construct the project in accordance with the contract documents

**General liability insurance:**   protects the contractor against claims from a third party for bodily injury or property damage

**Geotechnical report: also known as a soils report:**   A report prepared by a geotechnical engineering firm that includes the results of soil borings or test pits and recommends systems and procedures for foundations, roads, and excavation work

**Guaranteed maximum price contract:**   a type of cost-plus contract in which the contractor agrees to construct the project at or below a specified cost

**Income statement:**   a financial statement that summarizes the profitability of a company over a period of time

**Indirect construction costs:**   expenses indirectly incurred and not directly related to a specific project or construction activity, such as home office overhead

**Initial inspection:**   a quality-control inspection to ensure that workmanship and dimensional requirements are satisfactory

**Integrated project delivery:**   a delivery method in which a single relational contract is used to create a collaborative relationship among the project owner, designer, and the constructor

**Invitation to bid:**   a portion of the bidding documents soliciting bids for a project

**Job description:**   a description of the major tasks and duties to be performed by the person occupying a certain position

**Job hazard analysis:**   the process of identifying all hazards associated with a construction operation and selecting measures for eliminating, reducing, or responding to the hazards

**Jobsite general conditions costs:**   field indirect costs that cannot be tied to an item of work but are project-specific, and in the case of cost reimbursable contracts, considered part of the cost of the work

**Job specifications:**   the knowledge, abilities, and skills a person must possess to be able to perform the tasks and duties required

**Joint venture:**   a contractual collaboration of two or more parties to undertake a project

**Just-in-time delivery of materials:**   a material management philosophy in which supplies are delivered to the jobsite just in time to support construction activities; this minimizes the amount of space needed for on-site storage of materials.

**Labor and material payment bond:**   a surety instrument that guarantees that the contractor (or subcontractor) will make payments to his or her craftspeople, subcontractors, and suppliers

**Lean construction:**   process to improve costs incorporating efficient methods during both design and construction; includes value engineering and pull-planning

**LEED:** Leadership in Energy and Environmental Design, a system for evaluating the sustainability of a project; the system is administered by the United States Green Building Council.

**Letter of intent:** a letter, in lieu of a contract, notifying the contractor that the owner intends to enter into a contract pending resolution of some restraining factors, such as permits or financing

**Lien:** a legal encumbrance against real or financial property for work, material, or services rendered to add value to that property

**Lien release:** a document signed by a subcontractor or the general contractor releasing its rights to place a lien on the project

**Life-cycle cost:** the sum of all acquisition, operation, maintenance, use, and disposal costs for a product over its useful life

**Liquidated damages:** an amount specified in the contract that is owed by the contractor to the owner as compensation for damaged incurred as a result of the contractor's failure to complete the project by the date specified in the contract

**Litigation:** a court process for resolving disputes

**Long-form purchase order:** a contract for the acquisition of materials that is used by the project manager or the construction firm's purchasing department to procure major materials for a project

**Lump sum contract: also known as fixed-price or stipulated-sum contract:** A contract that provides a specific price for a defined scope of work

**Mark-up:** percentage added to the direct cost of the work to cover such items as overhead, fee, taxes, and insurance

**MasterFormat:** a 49-division numerical system of organization developed by the Construction Specifications Institute that is used to organize contract specifications and cost estimates

**Materialman's notice:** a notice sent to the owner as notice that the supplier will be delivering materials to the project

**Material supplier:** vendor who provides materials but no on-site craft labor

**Mediation:** a method of resolving disputes in which a neutral mediator is used to facilitate negotiations between the parties to the dispute

**Meeting agenda:** a sequential listing of topics to be addressed in a meeting

**Meeting notes:** a written record of meeting attendees, topics addressed, decisions made, open issues, and responsibilities for open issues

**Meeting notice:** a written announcement of a meeting, generally contains the date, time, and location of the meeting as well as the topics to be addressed

**Mock-ups:** stand-alone samples of completed work, such as a 6-foot-by-6-foot sample of a brick wall

**NanoEngineering Building:** new research laboratory and classroom building on the University of Washington campus in Seattle, Washington which served as the case study project throughout this edition of the text

**Network diagrams:** schedule that shows the relationships among the project activities with a series of nodes and connecting lines

**Notice to proceed:** written communication issued by the owner to the contractor, authorizing the contractor to proceed with the project and establishing the date for project commencement

**Occupational Safety and Health Administration:**   federal agency responsible for establishing jobsite safety standards and enforcing them through inspection of construction work sites

**Officer-in-charge:**   general contractor's principal individual who supervises the project manager and is responsible for overall contract compliance

**Offsite construction:**   prefabrication of building modules or systems improving on-site cost and schedule performance

**Open purchase order:**   a purchase order that specifies unit prices for specific materials, but does not specify an exact quantity for each material; may cover a specific period of time and contain a maximum total price. Delivery orders are made when the materials are needed, and invoices are provided with each delivery at the unit prices contained in the open purchase order.

**Operation and maintenance manuals:**   a collection of descriptive data needed by the owner to operate and maintain equipment installed on a project

**Over-billing:**   requesting payment for work that has not been completed

**Overhead:**   expenses incurred that do not directly relate to a specific project, for example, rent on the contractor's home office

**Overhead burden:**   a percentage mark-up that is applied to the total estimated direct cost of a project to cover overhead or indirect costs

**Partnering:**   a cooperative approach to project management that recognizes the importance of all members of the project team, establishes harmonious working relationships among team members, and resolves issues in a timely manner

**Partnering charter:**   documents the results of the partnering workshop; contains the project team's mission, identifies their collective goals, and is signed by workshop participants

**Payment bond:**   see *labor and material payment bond*

**Performance appraisal:**   a written evaluation of an individual's work performance during a specific review period

**Performance bond:**   a surety instrument that guarantees that the contractor will complete the project in accordance with the contract and protects the owner from the general contractor's default and the general contractor from the subcontractor's default

**Performance standards:**   standards a person is expected to achieve in the performance of his or her job

**Phased project-specific accident prevention plan:**   a detailed accident prevention plan that is focused directly on the hazards that will exist on a specific project and on measures that can be taken to reduce the likelihood of accidents

**Plugs:**   general contractor's cost estimates for subcontracted scopes of work

**Post-project analysis:**   a review of all aspects of the completed project to determine lessons that can be applied to future projects

**Pre-bid conference:**   meeting of bidding contractors with the project owner and architect to explain the project and bid process and solicit questions regarding the design or contract requirements

**Precedence diagramming method:**   scheduling technique that uses nodes to depict activities and arrows to depict relationships among the activities; used by most scheduling software

**Preconstruction agreement:**   a short contract that describes the contractor's responsibilities and compensation for preconstruction services

**Preconstruction conference:**   meeting conducted by owner or designer to introduce project participants and to discuss project issues and management procedures

**Preconstruction services:**   services that a construction contractor performs for a project owner during design development and before construction starts

**Pre-final inspection:**   an inspection conducted when the project is near completion to identify all work that needs to be completed or corrected before the project can be considered completed

**Preparatory inspection:**   a quality-control inspection to ensure that all preliminary work has been completed on a project site before starting the next phase of work

**Pre-proposal conference:**   meeting of potential contractors with the project owner and architect to explain the project, the negotiating process, and selection criteria and solicit questions regarding the design or contract requirements

**Prequalification of contractors:**   investigating and evaluating prospective contractors based on selected criteria prior to inviting them to submit bids or proposals

**Product data sheet: also known as material data or cut sheets:**   Information furnished by a manufacturer to illustrate a material, product, or system for some portion of the project which includes illustrations, standard schedules, performance data, instructions, and warranty

**Profit:**   the contractor's net income after all expenses have been subtracted

**Progress payment requests:**   document or package of documents requesting progress payments for work performed during the period covered by the request, usually monthly

**Progress payments:**   periodic (usually monthly) payments made during the course of a construction project to cover the value of work satisfactorily completed during the previous period

**Project close-out:**   completing the physical construction of the project, submitting all required documentation to the owner, and financially closing out the project

**Project control:**   the methods a project manager uses to anticipate, monitor, and adjust to risks and trends in controlling costs and schedules

**Project engineer:**   project management team member who assists the project manager on larger projects; more experienced and has more responsibilities than the field engineer but less than the project manager; responsible for management of technical issues on the jobsite

**Project labor curve:**   a plot of estimated labor-hours required per month for the duration of the project

**Project management:**   application of knowledge, skills, tools, and techniques to the many activities necessary to complete a project successfully

**Project management organization:**   the contractor's project management group headed by the officer-in-charge, including field supervision and staff

**Project manager:**   the leader of the contractor's project team who is responsible for ensuring that all contract requirements are achieved safely and within the desired budget and time frame

**Project manual:**   a volume usually containing the instructions to bidders, the bid form, general conditions, and special conditions; may also include a geotechnical report

**Project planning:**   the process of selecting the construction methods and the sequence of work to be used on a project

**Project start-up:**   mobilizing the project management team, establishing the project management office, and creating project document management systems

**Project team:**   individuals from one or several organizations who work together as a cohesive team to construct a project

**Project team list:**   list of all team members with their addresses and telephone contact information

**Property damage insurance:**   protects the contractor against financial loss due to damage to the contractor's property

**Public–private partnership:**   a project delivery method in which a private sponsor executes a contract with a public agency to finance, design, build, and operate a project for a specified period of time

**Pull-Planning:**   scheduling method often utilizing sticky-notes where milestones of each design or construction discipline are established and the project is scheduled backward with the aid of short-term detailed schedules

**Punch list:**   a list of items that need to be corrected or completed before the project can be considered completed

**Purchase orders:**   written contracts for the purchase of materials and equipment from suppliers

**Quality control:**   process to ensure materials and installation meet or exceed the requirements of the contract documents

**Quantity take-off:**   one of the first steps in the estimating process to measure and count items of work to which unit prices will later be applied to determine a project cost estimate

**Reading file:**   correspondence gathered in increments, for example, daily or weekly, routed among project staff, and then filed

**Reimbursable costs:**   costs incurred on a project that are reimbursed by the owner. The categories of costs that are reimbursable are specifically stated in the contract agreement.

**Request for information:**   document used to clarify discrepancies between differing contract documents and between assumed and actual field conditions

**Request for proposals:**   document containing instructions to prospective contractors regarding documentation required and the process to be used in selecting the contractor for a project

**Request for qualifications:**   a request for prospective contractors or subcontractors to submit a specific set of documents to demonstrate the firm's qualifications for a specific project

**Request for quotation:**   a request for a prospective subcontractor to submit a quotation for a defined scope of work

**Retention: also known as retainage:**   A portion withheld from progress payments for contractors and subcontractors to create an account for finishing the work of any parties not able to or unwilling to do so

**Rough order-of-magnitude cost estimate:**   a conceptual cost estimate usually based on the size of the project; prepared early in the estimating process to establish a preliminary budget and decide whether to pursue the project

**Safety data sheets:**   short technical reports that identify all known hazards associated with particular materials and provide procedures for using, handling, and storing the materials safely

**Schedule of submittals:**   a listing of all submittals required by the contract specifications

**Schedule of values:**   an allocation of the entire project cost to each of the various work packages required to complete a project and used to develop a cash flow curve for an owner and to support requests for progress payments

**Schedule update:**   schedule revision to reflect the actual time spent on each activity to date

**Self-performed work:**   project work performed by the general contractor's work force rather than by a subcontractor

**Shop drawing:** drawing prepared by a contractor, subcontractor, vendor, or manufacturer to illustrate construction materials, dimensions, installation, or other information relating to the incorporation of the items into a construction project

**Short-form purchase order:** purchase orders used on project sites by superintendents to order materials from local suppliers

**Short-interval schedule:** schedule that lists the activities to be completed during a short interval (2–4 weeks); also known as *look-ahead schedule*; used by the superintendent and foremen to manage the work

**Site logistics plan:** pre-project planning tool created often by the general contractor's superintendent, incorporates several elements including temporary storm water control, hoisting locations, parking, trailer locations, fences, traffic plans, and the like

**Soils report:** see *geotechnical report*

**Special conditions: also known as supplementary conditions:** A part of the construction contract that supplements and may also modify, add to, or delete portions of the general conditions

**Specialty contractors:** construction firms that specialize in specific areas of construction work, such as painting, roofing, or mechanical; such firms typically are involved in construction projects as subcontractors.

**Start-up checklist:** a listing of items that should be completed during project start-up, together with the date of completion for each item

**Start-up log:** a spreadsheet used to manage start-up activities relating to suppliers and subcontractors

**Storm water pollution prevention plan:** process to control storm water runoff often permitted, administered and inspected by the city or state

**Subcontractors:** specialty contractors who contract with and are under the supervision of the general contractor

**Subcontractor call sheet:** a form used to list all of the bidding firms from which the general contractor is soliciting subcontractor and vendor quotations

**Subcontractor preconstruction meeting:** a meeting the project manager and/or superintendent conduct with each subcontractor before allowing him or her to start work on a project

**Subcontracts:** written contracts between the general contractor and specialty contractors who provide craft labor and usually material for specialized areas of work

**Submittal register:** see *schedule of submittals*

**Submittals:** shop drawings, product data sheets, and samples submitted by contractors and subcontractors for verification by the design team that the materials purchased for installation comply with the design intent

**Substantial completion:** state of a project when it is sufficiently completed that the owner can use it for its intended purpose

**Summary schedule:** abbreviated version of a detailed construction schedule that may include 10 to 20 major activities

**Superintendent:** individual from the contractor's project team who is the leader on the jobsite and who is responsible for supervision of daily field operations on the project

**Surety:** a company that provides a bond guaranteeing that another company will perform in accordance with the terms of a contract

**Sustainability:**   broad term incorporating many green-building design and construction goals and processes including LEED

**Technical specifications:**   a part of the construction contract that provides the qualitative requirements for a project in terms of materials, equipment, and workmanship

**Telephone memorandum:**   a written summary of a telephone conversation

**Third-tier subcontractor:**   a subcontractor who is hired by a firm that has a subcontract with the general contractor

**Time-and-materials contract:**   a cost-plus contract in which the owner and the contractor agree to a labor rate that includes the contractor's profit and overhead. Reimbursement to the contractor is made based on the actual costs for materials and the agreed labor rate times the number of hours worked.

**Traditional project delivery method:**   a delivery method in which the owner has a contract with an architect to prepare a design for a project. When the design is completed, the owner hires a contractor to construct the project.

**Transmittal:**   a form used as a cover sheet for formally transmitting documents between parties

**Umbrella liability insurance:**   provides coverage against liability claims exceeding that covered by standard general liability or automobile insurance

**Unconditional lien release:**   a lien release indicating that the issuer has received a certain amount of payment and releases all lien rights associated with that amount

**Unit price contract:**   a contract that contains an estimated quantity for each element of work and a unit price, the actual cost determined once the work is completed and the total quantity of work is measured

**Value engineering:**   a study of the relative value of various materials and construction techniques to identify the least costly alternative without sacrificing quality or performance

**Warranty:**   a guarantee that all materials furnished are new and able to perform as specified and that all work is free from defects in material or workmanship

**Warranty service log:**   a spread sheet used to track all warranty claims from receipt to resolution

**Work breakdown structure:**   a list of significant work items that will have associated cost or schedule implications

**Worker's compensation insurance:**   protects the contractor from a claim due to injury or death of an employee on the project site

**Work package:**   a defined segment of the work required to complete a project

# Index